BAEN BOOKS by CHARLES SHEFFIELD:

BORDERLANDS OF SCIENCE

How to Think Like a Scientist and Write Science Fiction

CHARLES SHEFFIELD

BORDERLANDS OF SCIENCE

A Baen Books Original

Baen Publishing Enterprises
P.O. Box 1403
Riverdale, NY 10471
www.baen.com

ISBN: 0-671-31953-1

Cover art by Patrick Turner

First paperback printing, November 2000

Library of Congress Cataloging-in-Publication No. 99-37427

Distributed by Simon & Schuster
1230 Avenue of the Americas
New York, NY 10020

Production by Windhaven Press, Auburn, NH
Printed in the United States of America

THANKS

This book grew partly from a one-week series of seminars given at Dixie College, Utah, in the spring of 1998, on the subject of science and science fiction. I would like to thank all the participants in those seminars, particularly Ace Pilkington who organized and master-minded the event, for their useful analysis and comment. Before I went to Utah, I sent my material to Jim Baen asking him to look at what I had, and suggest places where the arguments seemed weak or insufficient. Much of the material had originally appeared in magazines that he published, and he offered considerable feedback. Several sections of the book owe much to our e-mail discussions. He also suggested that the material might make a useful and informative book.

Finally I would like to thank Nancy Kress and Joel Welling, who read everything that I wrote and told me when I was unintelligible because of brevity, or in danger of going overboard on technical detail. I usually took their advice, but not always, so any residual incomprehensibility is due to me alone.

Parts of this book are drawn from articles in *New Destinies, Analog,* and the *Samsung Quarterly* magazines.

CONTENTS

INTRODUCTION 1

CHAPTER 1: The Borderlands of Science
1.1   What you are reading 5
1.2   Defining science fiction 7
1.3   The good, the bad, and the simply awful 7
1.4   What kind of writer? 12
1.5   Three times ten to the fourteenth furlongs per fortnight ... 18

CHAPTER 2:   The Realm of Physics
2.1   The small world: atoms and down 22
2.2   Quantum paradoxes 25
2.3   Quantum teleportation 27
2.4   Relativity 29
2.5   Special relativity 29
2.6   General relativity 33
2.7   Beyond the atom 34
2.8   Strange physics: superconductivity 39
2.9   Super properties 43
2.10 Meanwhile, electricity 46
2.11 Superconductivity and statistics 49
2.12 High-temperature superconductors 52
2.13 Making it work 54

CHAPTER 3: Physics in the Large
3.1   Stars 59
3.2   Stellar endings 64
3.3   Black holes 67

CHAPTER 4: Physics in the Very Large
4.1   Galaxies 75
4.2   The age of the universe 76
4.3   Early days 78
4.4   All the way back 81
4.5   The missing matter 84

4.6	The end of the universe	87
4.6	The Big Crunch	89
4.7	At the eschaton	90
4.8	Expansion forever	91
4.9	Life in the far future	93
4.10	Complications from the cosmological constant	96

CHAPTER 5: The Constraints of Chemistry

5.1	Isaac Asimov and the Timonium engine	101
5.2	The limits of strength	106
5.3	The production of energy	107
5.4	Organic and inorganic: building an alien	111
5.5	Building a horse	115
5.6	Fullerenes: a chemical surprise	117
5.7	A burning home: the oxygen planet	120

CHAPTER 6: The Limits of Biology

6.1	The miracle molecule	123
6.2	The mystery of sex	128
6.3	Viruses, RNA, prions, and the origin of life	133
6.4	Local, or universal? Life elsewhere	138
6.5	Aging and immortality	140
6.6	Aging: a second look	143
6.7	Tissue engineering	145

CHAPTER 7: New Worlds for Old

7.1	Mercury	151
7.2	Venus	153
7.3	Earth	156
7.4	The Moon	157
7.5	Mars	159
7.6	The moons of Mars	161
7.7	The asteroid belt	162
7.8	Jupiter	163
7.9	Saturn	168
7.10	Uranus	171
7.11	Neptune	173
7.12	Pluto and the limits of the solar system	176
7.13	Planets around other stars	178

CHAPTER 8: Spaceflight

8.1	The ways to space	191
8.2	Rocket spaceships	192
8.3	Measures of performance	195
8.4	Mass drivers	197
8.5	Ion rockets	198
8.6	Nuclear reactor rockets	198
8.7	Pulsed fission rockets	199
8.8	Pulsed fusion	200
8.9	Antimatter rockets	202
8.10	Photon rockets	204
8.11	Space travel without reaction mass	205
8.12	Gravity swingbys	205
8.13	Solar sails	207
8.14	Laser beam propulsion	208
8.15	The Bussard Ramjet	209
8.16	Hybrids	211
8.17	Laser-powered rockets	211
8.18	Ram Augmented Interstellar Rocket (RAIR)	212
8.19	The vacuum energy drive	212
8.20	Launch without rockets	213
8.21	The beanstalk	214
8.22	Theme and variations	219
8.23	The launch loop	220
8.24	Space colonies	221
8.25	Solar power satellites	223

CHAPTER 9: Far-out Alternatives

9.1	Problems of interstellar travel	229
9.2	The Search for Extraterrestrial Intelligence	230
9.3	The choice of signal carrier	232
9.4	Beam me up	238
9.5	A helping hand from relativity	240
9.6	Faster than light	241
9.7	Wormholes and loopholes	243

CHAPTER 10: *Deus ex Machina*

10.1	Computer limits	247

10.2	Biological computers	251
10.3	Quantum computers: making a virtue of necessity	258
10.4	Where are the robots?	260
10.5	Nanotechnology: the anything machine	262
10.6	Artificial life and assisted thought	264

CHAPTER 11: Chaos: The Unlicked Bear-Whelp

11.1	Chaos: new, or old?	267
11.2	How to become famous	268
11.3	Building models	270
11.4	Iterated functions	273
11.5	Sick curves and fractals	280
11.6	Strange attractors	284

CHAPTER 12: Future War

12.1	The Invisible Man	288
12.2	Death rays	290
12.3	The ultimate personal weapon	292
12.4	Cyborgs	295
12.5	Cleaning up after nuclear war	296

CHAPTER 13: Beyond Science

13.1	Scientific heresies	301
13.2	Dinosaur doom	302
13.3	Gaia: the Whole Earth Mother	305
13.4	Dr. Pauling and Vitamin C	308
13.5	Minds and machines	311
13.6	Diseases from space	315
13.7	Cold fusion	318
13.8	No Big Bang	322
13.9	Free energy	325
13.10	Wild powers	327
13.11	Beyond the edge of the world	329
13.12	One last heresy	331

CHAPTER 14: The End of Science ... 339

APPENDIX: SCIENCE BITES ... 345

REFERENCES ... 383

INDEX ... 389

INTRODUCTION

You are reading an out-of-date book. Since the frontiers of science constantly advance, today's discussion of the borderlands of science will be obsolete tomorrow.

Unfortunately, I can't tell you which parts of the book that you are holding need an immediate update. With luck, the short-term changes will be mostly in the details, and the roots of each subject covered will survive intact. The biggest changes in science usually turn out to be the most surprising, the least predicted, and the slowest to be accepted.

Around the year 1900 there were plenty of forecasts as to what the coming century might bring. They were all wrong, not because of what they included, but because of what they left out. On the abstract side, no one expected relativity, quantum theory, the expansion of the universe, holography, subatomic structure, the conversion of matter to energy, solid state physics devices (such as transistors), information theory, black holes, the molecular structure of DNA, retroviruses, genome mapping, and the theory of finite state automata. Still less did anyone expect the torrent of practical applications, with their massive social fallout, that would follow from the new theories: television and telephones in almost every home, personal computers, supersonic aircraft, humans to the Moon and observing equipment to the planets, lasers, genetic engineering, video recorders, antibiotics, CAT scans, nuclear energy plants and nuclear bombs, and artificial satellites in regular use for communications, weather, and monitoring of the Earth's surface. No one in 1900 imagined that by 2000 the automobile would be absolutely central to many people's lives, as the principal means of transportation, recreation, and even courtship. Even in 1950, not a person on the planet would have predicted the existence of hundreds of millions of computers, used daily to conduct business, play games, send and receive mail, and wander at will through a world-wide information network.

Given our track record, and the fact that changes seem to be ever faster and more confusing, a pessimist could conclude that it is now impossible to write science fiction. Prediction of future conditions is impossible; even if we get the science right, surely the consequent social changes will be nothing like we suppose or can suppose. When reality is so surprising, what place is there for imagined worlds?

I prefer to argue as an optimist. In science fiction, new science and new applications mean an endless supply of new story ideas. So long as science and technology continue to advance, we can never run out of subject matter.

This book should be regarded as a beginning, not an end. It defines the frontiers—"borderlands"—of today's science. Those frontiers are not fixed, but constantly expanding. As they expand, the territory just beyond them comes into view. In that territory, waiting to be picked up and used, lie hundreds and thousands of gorgeous story ideas. They are pristine ideas, never used before, because they sit on ground never before explored.

I invite you to join me in wandering the new territories, picking up the best ideas, and using them. My idea is to offer a starting place for the exploration, but certainly not an end point. For one thing this book, like any book ever written, reflects the author's personal interests and obsessions. It's not reasonable that your own favorite scenery will exactly match mine.

A couple of final points need to be made. The first is an answer to the natural question: Why, with the Internet an integral part of most people's lives, do I not direct the reader to web sites for information? One answer will be obvious to you if you happen to be reading this book at the beach, or, as I like to think, secretly in a classroom while a teacher spouts New Age non-science at you. A book provides easy access, unobtrusively and with no need of special equipment.

Second, if you want to talk about lack of quality control there is no better example than the Internet. Normally, if it comes from somewhere like the Jet Propulsion Lab or the National Institutes of Health, it should be reliable. But even that is not safe. Some names of well-known individuals and institutions have been pre-empted as web site names. You think you are reading a report from the famous Dr. X. What you don't know is that Dr. X. is at this very moment engaged in a lawsuit regarding the theft of his good name and reputation.

One more caveat. This book makes another assumption. As a friend of mine, Roger Allen, said to me, "You call it *science* fiction, whereas most people pronounce it as science *fiction*." I plead guilty. That is indeed the way I view the science fiction field, or at least the part of the field that interests me. I assume that you, the reader, are interested in reading (and possibly writing) science fiction stories with some reasonable emphasis on science. If not, then this is not the book for you.

CHAPTER 1
The Borderlands of Science

1.1 What you are reading. This is a text for the writer or critical reader who likes the science of stories to be right. We will define the limits of knowledge in many areas, then wander beyond them. We will spend little time surveying the scientific mainstream. Many other books do that, taking a detailed look at quantum theory, astronomy, spaceflight, genetics, chemistry, or any other science you care to mention. We will offer the brief summaries that we need, and list some of the better reference works. Then we'll head for the scientific outer limits.

We will not try to tell you *how to write*. Nothing here will address plot, character, pacing, or style; nowhere will you see anything about markets, or foreign rights, or literary agents. When backgrounds appear, it will be for their scientific content only.

Plenty of other works address the problems of being a writer, discussing everything from style to contract negotiation to royalty rates. There are also writing courses without number. These courses are valuable, especially when taught by successful writers, but not one of the courses—even when they are explicitly and specifically about science fiction—teaches anything about science. We, by contrast, will be concerned with only one thing: making the science in stories accurate, current, plausible (if the story is set in the future), and *interesting*. Readers

5

of science fiction are an enthusiastic and forgiving audience. A writer of science fiction can perpetrate literary sins that are anathema in "mainstream" writing. But one thing you cannot get away with in my universe is botching the science of your story.

Or rather, you may get away with it some of the time. Your editors, who usually have a literary background but often lack a science background, may not catch you. Your readers will. Write about Shakespeare's *Paradise Lost*, or say that Abraham Lincoln led America in the Revolutionary War, and the editor will jump all over you. Claim that Titan is a moon of Jupiter, and nothing may be said. I did not make up this example. It happened. Titan as a moon of Jupiter sailed right past the editor and past the copy editor. A reader totally outside the book's production process (me) caught the blunder, and it was corrected in the published work. But you cannot rely on friendly readers being around all the time.

If you wander wildly beyond what scientists believe theoretically possible, you have to explain how and why. And you have to be reasonably current in your knowledge, because science changes constantly, and sometimes it changes fast. Three years ago, the idea of life anywhere in the universe, except on Earth, was pure speculation; today there is evidence, much disputed, for early life-forms on Mars.

As Josh Billings put it, *"It's not what we don't know that causes the trouble, it's the things we know that ain't so."*

Not all sciences are addressed in this book. When a field is omitted, one or more of the following will apply:

1) The topic doesn't seem to me to provide good material for science fiction stories.

2) Some other popular text covers the ground thoroughly and well.

3) I do not feel qualified to discuss the subject.

4) I do not believe that the subject, regardless of the fact that it may use the word "science" in its name, is real science.

A number of fringe areas, useful for stories whether or not you believe the theories, are described in Chapter 13.

1.2 Defining science fiction. When science fiction writers and readers get together, one of the things they are likely to talk about is the definition of science fiction. It's hard to reach agreement. I have my own definition, which, if it has no other virtue, describes the sort of science fiction that I like to read and write. It takes a few sentences and needs a brief preamble, but the definition goes as follows:

Science forms a great, sprawling continent, a body of learning and theories. Everything in science is interconnected, however loosely. If your theory doesn't connect with any part of the rest of science, you may be a genius with a new and profound understanding of the universe; but chances are you're wrong.

Science fiction consists of stories set on the shore or out in the shallow coastal water of that huge scientific land mass. Stay inland, safe above high tide, and your story will be not science fiction, but fiction about science. Stray too far, out of sight of land, and you are in danger of writing fantasy—even if you think it's science fiction.

The purpose of this book is to define the boundaries of science. Where do the limits lie, today, that define the scientific leading edge? And can we see places where, although no land is visible, prevailing currents or the sight of breakers convince us that it must exist? That, surely, is where we will find fertile ground for science fiction. On the other hand, we don't want to find ourselves out of our depth.

1.3 The good, the bad, and the simply awful: an example.

That's probably more than enough metaphors. Let me illustrate my point with a particular case.

Suppose I decide to write a story that tells of a race of alien beings who come to Earth from a home world orbiting the star Rigel. Their ships are enormous and fast—they are five miles long, and they can travel at 5,000 miles a second. When the aliens land on Earth and march out of their ships, it turns out that they are also huge; they are a hundred feet high and two hundred across, and they look, breed, and eat just like giant spiders.

Why are they here? To befriend humans, to educate us, to bring us into the Galactic federation of races, to enslave us, or to kill us?

One of their leaders explains to our representative. They are an ancient species, with a recorded history going back forty billion years. They were drawn to Earth by receipt of our radio signals, but humans, as primitive newcomers to the galaxy, are no more than food animals to them. They have come to overpower us, breed us, and eat us. At best, a few of us will be selected to help control the rest. As a reward, those humans who do cooperate will live a natural human life span.

Before our envoy can reply that the whole idea is intolerable, the Rigelian swallows him whole.

Humans seem doomed, until another brave earthling, a scientist, discovers that the aliens' eyes are different from ours. They see using shortwave ultraviolet light. We build a generator that can be used from miles away to beam an ultraviolet signal into the aliens' eyes. The repeating signal pattern interacts with the alien brain waves, sending them into convulsions and bringing them crashing to the ground. Humans approach and overpower them, learn the secret of the alien ship, and decide to go to Rigel and remove the alien menace from the galaxy forever.

An exceptionally dumb story? True. On the other hand, the smash-hit movie *Independence Day* was packed with worse scientific impossibilities and is in many ways a lot less plausible. I have never read anything quite like the tale I've described, but I will bet that the long-suffering editors of science fiction magazines see plenty.

What we have here is not science fiction, it is fantasy. Let's see why; and let's find out if we can, with a little juggling, convert it to science fiction.

First, consider how the aliens got here. A ship that travels at 5,000 miles a second (8,000 kilometers a second) sounds fast, but Rigel is more than 500 light-years away from the Sun. Light travels at almost 300,000 kilometers a second. Our aliens must be awfully patient in waiting for their dinners, because the journey here took them at least 18,000 years. If we intend to visit their home world and seek vengeance, it will take that long to get there.

The first fix: The aliens must possess some kind of faster-than-light (FTL) drive. They only use their "slow" drive at 5,000 miles a second when they are close to Earth. It is not necessary to specify how the FTL drive works. Science fiction has certain conventions, required by and used in so many stories that no explanation is called for. The FTL drive is one of them. If you want to create your own using the ideas of Chapter 9, that's fine. But you don't need to. Just say the aliens have one.

Next problem: The aliens supposedly had their attention drawn to Earth because they picked up our radio signals. But radio waves travel at the same speed as light, and we have been generating signals for only a century. Rigel is at least five hundred light-years away. The Rigelians ought not to know we even exist for another four hundred years or more.

The fix: In addition to the FTL drive, the aliens must possess a form of FTL communications system, able to pick up FTL emanations associated with normal electro-magnetic radiation. They knew of our presence as soon as we began to broadcast.

An alternative fix, to both this and the previous problem, might suggest itself to you: although the aliens came originally from Rigel, they have been colonizing space for a long time. Their nearest colony is much closer than Rigel. Unfortunately, this doesn't help unless the aliens were already closer to us than the nearest star. The travel time from Alpha Centauri at 5,000 miles a second is more than a century and a half—too long for them to get here after receiving our first radio signals.

The Rigelians are a hundred feet high. At that size, they would not be able to march out of their ships. In fact, they would not march anywhere, or even be able to move. The largest creature on Earth, the blue whale, is as much as a hundred feet long, but it is able to grow to such a size only because its body is supported by water. A land animal a hundred feet tall would lie like a beached whale, crushed by its own weight and unable to breathe. Weight increases as the cube of linear size, the area of an animal's limbs only as the square, so something fifteen times as tall as a human has to support fifteen times as much weight per square inch of limb cross-section.

Nothing made of living tissue and walking around on land today can be much bigger than an adult elephant, whose legs are short and thick. This is discussed in detail in J.B.S. Haldane's essay, "On Being The Right Size" (Haldane, 1927), which illustrates the size/area/weight relationship with a memorable image: "You can drop a mouse down a thousand-yard mine shaft; and, on arriving at the bottom, it gets a slight shock and walks away. A rat is killed, a man is broken, a horse splashes."

The fix: Our aliens must wear suits. Those suits include strong exoskeletons responding exactly to the movements that the alien within wishes to make. We could make a good approximation to such a suit today. It would be a trivial task for beings who build five-mile-long interstellar spaceships. The suits also avoid another big question: How do aliens happen to be able to breathe our air?

They came to eat us. This may seem psychologically improbable, but let us accept that we do not understand alien motives. There is a much bigger problem. It is highly unlikely that creatures who evolved independently of Earth life will have a body chemistry close enough to ours to be able to eat the same kinds of foods. Borrowing from Chapter 6, we note that amino acids are the raw materials from which proteins are produced; that every living thing on Earth produces amino acids using coded triplets of DNA or RNA bases; and that all codings produce only a total of twenty amino acids out of the hundreds possible. We might argue that alien body chemistry will also be based on something like RNA and DNA, because it is the only system we know of today for cell reproduction. However, the chance that the very same amino acids would be generated is remote. Human flesh would be more likely to poison aliens than be relished by them.

The fix: We call on some version of the *panspermia* theory (Chapter 13), according to which life on Earth did not develop independently, but was carried here from space. In that case, the aliens arose from the same space-borne seeds; therefore, they can have compatible body biochemistry and can digest humans.

An alternative version, in which an early super-race colonized the whole galaxy, then vanished without trace, can accomplish the same result.

More problems: the aliens see using shortwave ultra-violet light. It's not impossible for such eyes to have developed, particularly since the aliens come from a world whose blue-white sun, Rigel, produces far more energy in the UV than Sol. But the aliens' eyes are working here on Earth. Our atmosphere absorbs UV radiation of wavelengths shorter than 0.3 micrometers, so anything that used this part of the spectrum for vision would be blind on the surface of this planet.

We will not ask how our scientist learned enough about the alien brain to realize how to disable it. Presumably she is a genius in neuroscience and signal processing. But she doesn't know much basic physics. Her UV signal generator will not be effective at long distances, as she planned to use it, since the beam is strongly absorbed by air.

The fix: Give the aliens eyes that are superior to ours. Let them be sensitive to everything from short UV, at less than 0.25 micrometer wavelength, to reflective infrared at 2 micrometers. (We see from about 0.4 to 0.7 micrometers.) That way the human can see the light produced by her signal generator, and be a lot less nervous when using it in the field. ("Is this thing on?" "I don't see anything. Those aliens are getting awfully close.")

The science problems in this story were deliberately chosen as obvious. There are a couple of other, more subtle, scientific errors. They are minor, and one of them might get past a science fiction editor. Rather than discussing them here, I will let the reader discover the glitches and provide a fix.

Even when all the fantasy elements have been con-verted to science fiction, it remains a dumb story. But at least it is now a stupid *science fiction* story.

Do you really need to worry about any of this? After all, what you are going to write is fiction.

If you are writing the kind of science fiction that I want to read, you do. I have thrown books across the room and never picked them up again when they have offered some scientific howler too dumb to believe: a story in which a planet had an atmosphere of oxygen and hydrogen (watch out for that cigarette!); a man with an IQ of 5,000 (measured how?); a scientist, warning the President that an eclipse of the galaxy is on the way.

Even some of the science fiction classics are guilty. Consider H.G. Wells' *The Invisible Man* (Wells, 1897). He took a drug which made his body of the same refractive index as air. But if your eyes did not absorb light, you would be blind.

There are more plausible ways of making a person invisible. Think about the problem and see what you can come up with. We will address the topic in more detail in Chapter 12.

1.4 What kind of writer? You are going to be a science fiction writer. What kind?

It's no good saying, "a rich and famous one." We need to be more specific. Let's look at the options, going back to the roots of science fiction.

When science fiction writers get together, one of the things they talk about is *when science fiction began*.

Was it with Lucian of Samosata, who almost two thousand years ago had his hero, Icaromenippus, wonder about the moon, and then go there? Even farther back, is Homer's *Odyssey* science fiction?

What about Marco Polo's travels to China, which some today say never happened, with the tale made up in a prison cell? How about the voyages of Sinbad? The idea of a bird that could carry off and dine on elephants must have seemed more probable to early Europeans than the elephant itself—surely a fine candidate for a mythical beast.

Closer to our own time, we have Kepler's *Somnium* (Kepler, 1634) and Francis Godwin's *The Man in the Moone* (Godwin, 1638). And what about *Gulliver's*

Travels, travel not to the Moon but to places just as strange?

These were all seventeenth- and eighteenth-century works, although only Swift's stories are well-known today. If you are inclined to dismiss them as social satires, let me point out that in the "Voyage to Laputa," published in 1726, Swift remarked that the astronomers of that flying island, with their superior telescopes, had "discovered two lesser stars, or 'satellites,' which revolve around about Mars, whereof the innermost is distant from the center of the primary planet exactly three of its diameters, and the outermost, five; the former revolves in the space of ten hours and the latter in twenty-one and a half."

The modern values of these numbers are 1.35 Mars-diameters and 7 hrs. 39 mins. for Phobos, and 3.5 diameters and 30 hrs. 18 mins. for Deimos. However, no one knew that Mars had moons at all until Asaph Hall discovered Deimos and Phobos in 1877.

Inspired prediction? More probably, sheer coincidence— but any modern science fiction writer would be proud to do as well.

All these works have been cited as the "first science fiction." However, they all form isolated data points. They did not give rise to a "school" of writers, near-contemporaries who went on to write similar works. There was no continuity with what came after them.

That continuity came with two nineteenth-century authors. The different types of story that they created persist. And the differences are relevant to writers today.

Origin number one was Mary Shelley's *Frankenstein* (Shelley, 1818).

This novel is significant in three ways. First, it has been continuously of interest—and read—since it was written in 1816. Second, it is fiction *about science*. The idea of the reanimation of a corpse was based on the science of the day. Mary Shelley wrote after the experiments of Galvani and Franklin, but before Michael Faraday and James Clerk Maxwell put an understanding of electricity and magnetism on a firmer footing. At the time when *Frankenstein* was written, electricity was perhaps the biggest mystery of the day. Even such authorities as Erasmus Darwin, the celebrated grandfather of Charles Darwin, did not dismiss the idea of the spontaneous generation of life. How much easier, then, it must have seemed to reanimate a corpse, rather than to create life from inanimate materials. Mary Shelley was offering *legitimate scientific speculation*, not fantasy.

Third, *Frankenstein* is a *moral* tale, concerned ultimately less with science than with moral issues.

You might not think about moral questions if you are familiar only with the movie versions of the story. There the monster is at center stage. The cinematic electrical effects and the harnessing of the lightning are not in the original book at all. In the movies, Baron Frankenstein loses ground to his creation. Even his name is abused. When children say, "you look like Frankenstein," we all know what they mean, while in that classic work, *Abbott and Costello Meet Frankenstein*, the Baron has gone entirely, replaced by Count Dracula (who loses out in the end to the Wolf Man).

Let me add one personal anecdote about the power of the story. My first exposure was neither to book nor to movie. The tale was *told* to me by my father when I was about eight, and I realize now that he talked of the movie version. It fascinated and absolutely terrified me.

The second point of origin is Jules Verne. Between 1860 and 1870, he wrote *From the Earth to the Moon, Journey to the Center of the Earth, Five Weeks in a Balloon*, and *Twenty Thousand Leagues Beneath the Sea*.

Like *Frankenstein*, these books have been continuously read since the day that they were published. They also display a great interest in science. (Verne himself grumbled, late in life, that the upstart H.G. Wells didn't use sound scientific methods, the way that Verne had.)

Verne, however, is little concerned with moral issues. In contrast with Mary Shelley's story, Verne's plots can be summarized as "a bunch of cheerful but emotionally challenged guys go off and have a rattling (and scientific) good time."

There is one other way in which Mary Shelley and Jules Verne differ profoundly. Verne's work directly influenced scientists—not of his own generation, but of the one that followed. I don't think anyone reading *Frankenstein* around 1825 said, "Hey, let's go and collect a few bits of dead bodies and see what we can do." But we know for a fact that Tsiolkovsky, the father of the Russian space program, was inspired by Verne. Hermann Oberth, whose work in turn inspired Wernher von Braun, discovered Verne's *From the Earth to the Moon* when he was eleven years old, and was led to a lifelong commitment to space flight. Finally, Robert Goddard was influenced by H.G. Wells, and *The War of the Worlds*. He later said that after reading it, "I imagined how wonderful it would be to make some device which had even the *possibility* of ascending to Mars, and how it would look on a small scale if sent up from the meadow at my feet. . . . Existence at last seemed very purposive."

Today, the lines of descent from Mary Shelley and Jules Verne have converged. We all want to write stories that draw from both—morally significant tales, with first-rate science.

Given that overall desire, we have a number of options as to the type of writer we want to be. I will name and define half a dozen categories: Bandwagoners, Bards, Importers, Seers, Sensitives, and World-builders. And I will make my own recommendation as to how a new writer should proceed.

The Bandwagoner. This writer is very much of the moment. The story chooses a theme, often of current social or scientific significance, and pushes it hard. You will like the result if you are on the same intellectual wavelength, and share the same passion. Otherwise . . .

It is difficult to be consistently successful with this kind of writing. Causes change with the times. Not only that, the writer is likely to compete with fifty others who have responded to the same hot topic. Finally, the Bandwagoner must find an editor who shares the same point of view.

The Bard. If you can visualize an interesting character passing through a whole succession of intriguing situations, and can describe it so that other people can see the same scenes, you qualify as a Bard. If you put your character in space, you are likely to be writing space opera. If she is in Neverland, you are writing fantasy. If he is wandering the Mediterranean, a long time ago, someone else has already done the story.

Successful Bards have narrative strength, but more than that, they generate interest in the central character. Given a good tale-spinning talent, you can write these stories forever. Of course, after a while an acute reader may feel they are all the same story. Homer was smart, and wrote the *Odyssey* only once.

The Importer. Science fiction stories have been written about hundreds of branches of science. Fortunately for the Importer, science is a field in which new developments are reported every week. The Importer picks a subject, any subject: aardvarks to zygotes, and everything in between; quarks, game theory, prions, quasars, retroviruses, artificial life, superstrings.

There is only one restriction, but it's an important one: the subject must not have been used before in science fiction. The Importer learns enough to be convincing, then uses the subject as the background of a story.

This works better than one might suppose. Most science fiction readers have a natural interest in science, and a story is a painless way of acquiring new information. As a result, Importer stories can often be sold even when the writer violates some of the rules of good storytelling. The most common fault is in presenting science in big, lecture-like blocks—"expository lumps." The Importer gets away with it if the science is new enough, and interesting enough.

Of course, it does not have to be science. You can import historical eras, myths, systematic magic, or other branches of literature. Science has the advantage of continued development, and is therefore always a source of new material.

A good Importer imports, and then instinctively extrapolates. Many of the successful "predictions" of science fiction have come from Importers.

The Seer. If you can look at an everyday situation, and see it in a new and interesting perspective, you are a Seer. You are also a rarity. Writers of this type don't necessarily offer much in the way of plot or character, but they shed a new light on the world. If in reading a story you stop every few pages and say to yourself, "I never thought about it *that* way," then you have found yourself a first-class Seer.

How do you go about becoming one? I don't think you do. Unlike the Bandwagoner or Importer, which can be a learned skill, the Seer has an inborn talent, an inward eye that sees the world in a different way.

The Sensitive. It is an accident of timing or temperament that leads this writer to the science fiction field. The real interest of the Sensitive has nothing to do with science, it is in human (and nonhuman) emotions. The science elements of a Sensitive's stories are often nonexistent. When they are also present, the stories are the strongest in the field.

The World-builder. This writer is not particularly interested in the characters in the story, or even the plot. The fascination lies in the background—a planetary system, a future society, a well-designed alien, or an artificial world.

Science is important here. Many World-builders spend much time and effort making every element of their world consistent and plausible. It is necessary, too. Science fiction readers are *careful* readers. They are also communicative readers. Get something wrong, and you will hear about it.

Now for the promised recommendation. If you are a beginning writer, I suggest that you try to become an Importer. All it takes is the ability to read about a science subject, and then write about it, in the context of a story, in reasonably clear language.

Am I oversimplifying? A little, perhaps—maybe because I started out as an Importer myself. On the other hand, when I read some of my early published stories I am convinced that, other than as an Importer, there is no way that I would ever have seen myself in print.

1.5 Three times ten to the fourteenth furlongs per fortnight: units and notation.

Kilometers or miles? Kilograms or pounds? Knots, or miles per hour?

In principle, the United States operates using the metric system. In practice, it does not. There are also a variety of special measures used in science, such as the *fermi* (10^{-15} meters), *curie* (a unit of radioactivity), *dol* (a measure of pain), *flops* (in computers, the number of floating-point operations per second), *mho* (conductivity, the reciprocal of resistance—*ohm* backwards, get it?), and *barn* (a unit of area, 10^{-24} square meters—"It's as big as a barn," another good example of physicist humor).

There are also a number of would-be humorous units from other fields. The *millihelen* is defined, after Marlowe, as the amount of beauty sufficient to launch one ship. The *kan* is a suggested unit of modesty, of which the somewhat arrogant American scientist, Millikan, was said to possess one-thousandth.

Deliberate attempts at humor aside, what units should you use in a story?

The most familiar form is the best one.

Light-years are better than *parsecs*, even though astronomers almost always employ the latter. *Miles per hour*, rather than *knots* (nautical miles per hour), and *feet* rather than *fathoms*, unless you are specifically seeking a flavor of the sea—and even then, it's good to tie your numbers to something specific. "We'll never catch that sub, it must be doing over forty knots and heading down to a hundred fathoms."

Units for which the average reader has no instinct or training, such as the curie, dol, mho, and *gauss* (the measure of magnetic field), should be avoided completely. Kilometers are all right, and so, in general, are meters and kilograms. Even here, there are exceptions. When you watch the Olympic Games, do you, like me, have to convert the pole vault from meters to feet before you have a real idea of how high the bar is set? Do you know if a long jump of nine meters is poor, good, or a world record?

There is one golden rule: numbers are there not to prove how smart you are, but to provide information to the reader. Since this book is written for anyone who wants to use a reasonable amount of science in stories, it's not unfair of me to assume the same of my potential reader. But if you yourself have to sit down and work out how big something is in a particular unit, you should probably look for another way to get your point across.

Technical vocabulary, like the occasional number, adds a feeling of solidity to a story. It should be used sparingly. And get the term right, or don't use it at all. Do not say "quasar" if you mean "quark," confuse momentum with energy, or employ "light-year" as though it is a unit of time.

As for the notation in which very large or small numbers are written, in this book I will assume that you, the reader, are familiar with expressions such as 10^{+27}, or 5.3×10^{-16}. Even so, I advise you to avoid this form of notation whenever possible.

That point was brought home to me many years ago at, of all places, NASA headquarters. We were in the middle of a presentation, and casually throwing around expressions like 10^{12} and 10^{-8}, when a member of the audience (who happened to be NASA's Head of International Affairs) pointed at one of the numerical tables and said, "What are those little figures written above the tens?"

As we said when we left, "Thank God he's not designing the spacecraft."

Find an alternative in your storytelling to scientific notation. You never know whom you may lose when you use it.

CHAPTER 2
The Realm of Physics

Physics is the study of the properties of matter and energy. We begin with physics, not because it is easier, harder, or more important than other sciences, but because it is, in a specific sense, more fundamental.

More fundamental, in that from the laws of physics we can construct the laws of chemistry; from the laws of chemistry and the laws of physics together we can in turn build the laws of biology, of properties of materials, of meteorology, computer science, medicine, and anything else you care to mention. The process cannot be reversed. We cannot deduce the laws of physics from the laws of chemistry, or those of biology.

In practice, we have a long way to go. The properties of atoms and small molecules can be calculated completely, from first principles, using quantum theory. Large molecules present too big a computational problem, but it is considered to be just that, not a failure of understanding or principles. In the same way, although most biologists have faith in the fact that, by continuing effort, we will at last understand every aspect of living systems, we are a huge distance away from explaining things such as consciousness.

A number of scientists, such as Roger Penrose, believe that this will never happen, at least with current physical theories (Penrose, 1989, 1994; see also Chapter 13). Others, such as Marvin Minsky, strongly disagree; our brains are no more than "computers made of meat." Some scientists, believers in dualism, strongly disagree with that, asserting the existence of a basic element of mind quite divorced from the mechanical operations of the brain (Eccles, 1994).

Furthermore, there is a "more is different" school of scientists, led by physicist Philip Anderson and evolutionary biologist Ernst Mayr. Both argue (Anderson, 1972; Mayr, 1982) that one cannot deduce the properties of a large, complex assembly by analysis of its separate components. In Mayr's words, "the characteristics of the whole cannot (even in theory) be deduced from the most complete knowledge of the components, taken separately or in other partial combinations." For example, study of single cells would never allow one to predict that a suitable collection of those cells, which we happen to call the human brain, could develop self-consciousness.

Who is right? The debate goes on, with no end in sight. Meanwhile, this whole area forms a potential gold mine for writers.

2.1 The small world: atoms and down. It was Arthur Eddington who pointed out that, in size, we are slightly nearer to the atoms than the stars. It's a fairly close thing. I contain about 8×10^{27} atoms. The Sun would contain about 2.4×10^{28} of me. We will explore first the limits of the very small, and then the limits of the very large.

A hundred years ago, atoms were regarded as the ultimate, indivisible elements that make up the universe. That changed in a three-year period, when in quick succession Wilhelm Röntgen in 1895 discovered X-rays, in 1896 Henri Becquerel discovered radioactivity, and in 1897 J.J. Thomson discovered the electron. Each of these can only be explained by recognizing that atoms have an interior structure, and the behavior of matter and radiation in that sub-atomic world is very different from what we are used to for events on human scale.

The understanding of the micro-world took a time to appear, and it is peculiar indeed. In the words of Ilya Prigogine, a Nobel prize-winner in chemistry, "The quantum mechanics paradoxes can truly be said to be the nightmares of the classical mind."

The next step after Röntgen, Becquerel and Thomson came in 1900. Some rather specific questions as to how radiation should behave in an enclosure had arisen, questions that classical physics couldn't answer. Max Planck suggested a rather *ad hoc* assumption that the radiation was emitted and absorbed in discrete chunks, or *quanta* (singular, *quantum*; hence, a good deal later, *quantum theory*). Planck introduced a fundamental constant associated with the process. This is *Planck's constant*, denoted by h, and it is tiny. Its small size, compared with the energies, times, and masses of the events of everyday life, is the reason we are not aware of quantum effects all the time.

Most people thought that the Planck result was a gimmick, something that happened to give the right answer but did not represent anything either physical or of fundamental importance. That changed in 1905, when Albert Einstein used the idea of the quantum to explain another baffling result, the *photoelectric effect*.

Einstein suggested that light must be composed of particles called *photons*, each with a certain energy decided by the wavelength of the light. He published an equation relating the energy of light to its wavelength, and again Planck's constant, h, appeared. (It was for this work, rather than for the theory of relativity, that Einstein was awarded the 1921 Nobel Prize in physics. More on relativity later.)

While Einstein was analyzing the photoelectric effect, the New Zealand physicist Ernest Rutherford was studying the new phenomenon of radioactivity. The usual notion of the atom at the time was that of a sphere with electrical charges dotted about all over inside it, rather like raisins in a cake. Rutherford found his experiments were not consistent with such a model. Instead, an atom seemed to be made up of a very dense central region, the *nucleus*, surrounded by an orbiting cloud of electrons. In 1911 Rutherford proposed this new structure for the atom, and pointed out that while the atom itself was small—a few billionths of an inch—the nucleus was *tiny*, only about a hundred thousandth as big in radius as the whole atom. In other words, matter, everything from humans to stars, is mostly empty space and moving electric charges.

The next step was taken in 1913 by Niels Bohr. He applied the "quantization" idea of Planck and Einstein— the idea that things occur in discrete pieces, rather than continuous forms—to the structure of atoms proposed by Rutherford.

In the Bohr atom, electrons can only lose energy in chunks—quanta—rather than continuously. Thus they are permitted orbits only of certain energies, and when they move between orbits they emit or absorb radiation at specific wavelengths (light is a form of radiation, in the particular wavelength range that can be seen by human eyes). The electrons can't have intermediate positions, because to get there they would need to emit or absorb some fraction of a quantum of energy; by definition, fractions of quanta don't exist. The permitted energy losses in Bohr's theory were governed by the wavelengths of the emitted radiation, and again Planck's constant appeared in the formula.

It sounded crazy, but it worked. With his simple model, applied to the hydrogen atom, Bohr was able to calculate the right wavelengths of light emitted from hydrogen.

More progress came in 1923, when Louis de Broglie proposed that since Einstein had associated particles (photons) with light waves, wave properties ought to be assigned to particles such as electrons and protons. He tried it for the Bohr atom, and it worked.

The stage was set for the development of a complete form of quantum mechanics, one that would allow all the phenomena of the subatomic world to be tackled with a single theory. In 1925 Erwin Schrödinger employed the wave-particle duality of Einstein and de Broglie to come up with a basic equation that applied to almost all quantum mechanics problems; at the same time Werner Heisenberg, using the fact that atoms emit and absorb energy only in finite and well-determined pieces, produced another set of procedures that could also be applied to almost every problem.

Soon afterwards, in 1926, Paul Dirac, Carl Eckart, and Schrödinger himself showed that the Heisenberg and Schrödinger formulations can be viewed as two different approaches within one general framework. In 1928, Dirac took another important step, showing how to incorporate the effects of relativity into quantum theory.

It quickly became clear that the new theory of Heisenberg, Schrödinger, and Dirac allowed the internal structure of atoms and molecules to be calculated in detail. By 1930, quantum theory, or quantum mechanics as it was called, became *the* method for performing calculations in the world of molecules, atoms, and nuclear particles. It was the key to detailed chemical calculations, allowing Linus Pauling to declare, late in his long life, "I felt that by the end of 1930, or even the middle, that organic chemistry was pretty well taken care of, and inorganic chemistry and mineralogy— except the sulfide minerals, where even now more work needs to be done" (Horgan, 1996, p. 270).

2.2 Quantum paradoxes. Quantum theory was well-formulated by the end of the 1920s, but many of its mysteries persist to this day. One of the strangest of them, and the most fruitful in science fiction terms, is the famous paradox that has come to be known simply as "Schrödinger's cat." (We are giving here a highly abbreviated discussion. A good detailed survey of quantum theory, its history and its mysteries, can be found in the book *In Search of Schrodinger's Cat*; Gribbin, 1984.)

The cat paradox was published in 1935. Put a cat in a closed box, said Schrödinger, with a bottle of cyanide, a source of radioactivity, and a detector of radioactivity. Operate the detector for a period just long enough that there is a fifty-fifty chance that one radioactive decay will be recorded. If such a decay occurs, a mechanism crushes the cyanide bottle and the cat dies.

The question is: Without looking in the box, is the cat alive or dead? Quantum indeterminacy insists that until we open the box (i.e., perform the observation) the cat is partly in the two different states of being dead and being alive. Until we look inside, we have a cat that is neither alive nor dead, but half of each.

There are refinements of the same paradox, such as the one known as "Wigner's friend" (Eugene Wigner, born in 1902, was an outstanding Hungarian physicist in the middle of the action in the original development of quantum theory). In this version, the cat is replaced by a human being. That human being, as an observer, looks to see if the glass is broken, and therefore automatically removes the quantum indeterminacy. But suppose that we had a cat smart enough to do the same thing, and press a button? The variations—and the resulting debates—are endless.

With quantum indeterminacy comes uncertainty. *Heisenberg's uncertainty principle* asserts that we can never know both of certain pairs of variables precisely, and at the same time. Position and speed are two such variables. If we know exactly where an electron is located, we can't know its speed.

With quantum indeterminacy we also have the loss of another classical idea: *repeatability*. For example, an electron has two possible spins, which we will label as "spin up" and "spin down." The spin state is not established until we make an observation. Like Schrödinger's half dead/half alive cat, an electron can be half spin up and half spin down pending a measurement.

This has practical consequences. At the quantum level an experiment, repeated under what appear to be identical conditions, may not always give the same result. Measurement of the electron spin is a simple example, but the result is quite general. When we are dealing with the subatomic world, indeterminacy and lack of repeatability are as certain as death and taxes.

Notice that the situation is not, as you might think, merely a statement about our state of knowledge; i.e., we know that the spin is either up or down, but we don't know which. The spin is *up and down at the same time*. This may sound impossible, but quantum theory absolutely requires that such "mixed states" exist, and we can devise experiments which cannot be explained without mixed states. In these experiments, the separate parts of the mixed states can be made to interfere with each other.

To escape the philosophical problem of quantum indeterminacy (though not the practical one), Hugh Everett and John Wheeler in the 1950s offered an alternative "many-worlds theory" to resolve the paradox of Schrödinger's cat. The cat is both alive and dead, they say—but in different universes. Every time an observation is made, all possible outcomes occur. The universe splits at that point, one universe for each outcome. We see one result, because we live in only one universe. In another universe, the other outcome took place. This is true not only for cats in boxes, but for every other quantum phenomenon in which a mixed state is resolved by making a measurement. The change by measurement of a mixed state to a single defined state is often referred to as "collapsing the wave function."

An ingenious science fiction treatment of all this can be found in Frederik Pohl's novel *The Coming of the Quantum Cats* (Pohl, 1986).

Quantum theory has been defined since the 1920s as a computational tool; but its philosophical mysteries continue today. As Niels Bohr said of the subject, "If you think you understand it, that only shows you don't know the first thing about it."

To illustrate the continuing presence of mysteries, we consider something which could turn out to be the most important physical experiment of the century: the demonstration of *quantum teleportation*.

2.3 Quantum teleportation. Teleportation is an old idea in science fiction. A person steps into a booth here, and is instantly transported to another booth miles or possibly light-years away. It's a wonderfully attractive concept, especially to anyone who travels often by air.

Until 1998, the idea seemed like science fiction and nothing more. However, in October 1998 a paper was published in *Science* magazine with a decidedly science-fictional title: "Unconditional Quantum Teleportation." In that paper, the six authors describe the results of an experiment in which quantum teleportation was successfully demonstrated.

We have to delve a little into history to describe why the experiment was performed, and what its results mean. In 1935, Einstein, Podolsky, and Rosen published a "thought experiment" they had devised. Their objective was to show that something had to be wrong with quantum theory.

Consider, they said, a simple quantum system in which two particles are coupled together in one of their quantum variables. We will use as an example a pair of electrons, because we have already talked about electron spin. Einstein, Podolsky, and Rosen chose a different example, but the conclusions are the same.

Suppose that we have a pair of electrons, and we know that their total combined spin is zero. However, we have no idea of the spin of either individual electron, and according to quantum theory we cannot know this until we make an experiment. The experiment itself then forces an electron to a particular state, with spin up or spin down.

We allow the two electrons to separate, until they are an arbitrarily large distance apart. Now we make an observation of one of the electrons. It is forced into a particular spin state. However, since the total spin of the pair was zero, the other electron must go into the opposite spin state. This happens at once, no matter how far apart the electrons may be.

Since nothing—including a signal—can travel faster through space than the speed of light, Einstein, Podolsky, and Rosen concluded that there must be something wrong with quantum theory.

Actually, the thought experiment leads to one of two alternative conclusions. *Either* there is something wrong with quantum theory, *or* the universe is "nonlocal" and distant events can be coupled by something other than signals traveling at or less than the speed of light.

It turns out that Einstein, Podolsky, and Rosen, seeking to undermine quantum theory, offered the first line of logic by which the locality or nonlocality of the universe can be explored; and experiments, first performed in the 1970s, came down in favor of quantum theory and a nonlocal universe. Objects, such as pairs of electrons, can be "entangled" at the quantum level, in such a way that something done to one *instantaneously* affects the other. This is true in principle if the electrons are close together, or light-years apart.

To this time, the most common reaction to the experiments demonstrating nonlocality has been to say, "All right. You can force action at a distance using 'entangled' particle pairs; but you can't make use of this to send information." The new experiment shows that this is not the case. Quantum states were transported (teleported) and information was transferred.

The initial experiment did not operate over large distances. It is not clear how far this technique can be advanced, or what practical limits there may be on quantum entanglement (coupled states tend to decouple from each other, because of their interactions with the rest of the universe). However, at the very least, these results are fascinating. At most, this may be the first crack in the iron straitjacket of relativity, the prodigiously productive theory which has assured us for most of the 20th century that faster-than-light transportation is impossible.

We now consider relativity and its implications.

2.4 Relativity. The second great physical theory of the twentieth century, as important to our understanding of Nature as quantum theory, is relativity. Actually, there are in a sense *two* theories of relativity: the special theory, published by Einstein in 1905, and the general theory, published by him in 1915.

2.5 Special relativity. The special theory of relativity concentrates on objects that move relative to each other at constant velocity. The general theory allows objects to be accelerated relative to each other in any way, and it includes a theory of gravity.

Relativity is often thought to be a "hard" subject. It really isn't, although the general theory calls for a good deal of mathematics. What relativity is, more than anything, is *unfamiliar*. Before the effects of relativity are noticed, things need to be moving relative to each other very fast (a substantial fraction of the speed of light), or they must involve a very strong gravitational field. We are as unaware of relativity as a moving snail is unaware of wind resistance, and for the same reason; our everyday speeds of motion are too slow for the effects to be noticed.

Everyone from Einstein himself to Bertrand Russell has written popular accounts of relativity. We name just half a dozen references, in increasing order of difficulty: *Einstein's Universe* (Calder, 1979); *The Riddle of Gravitation* (Bergmann, 1968); *Relativity and Common Sense* (Bondi, 1964); *Einstein's Theory of Relativity* (Born, 1924); *The Meaning of Relativity* (Einstein, 1953); and *Theory of Relativity* (Pauli, 1956). Rather than talk about the theory itself, we are going to confine ourselves here to its major *consequences*. In the case of special relativity, there are six main ones to notice and remember.

1) Mass and energy are not independent quantities, but can be converted to each other. The formula relating the two is the famous $E = mc^2$.

2) Time is not an absolute measure, the same for all observers. Instead, time passes more slowly on a moving object than it does relative to an observer of that object. The rule is, for an object traveling at a fraction F of the speed of light, when an interval T passes onboard the object, an interval of $1/\sqrt{(1-F^2)}$ of T passes for the observer. For example, if a spaceship passes you traveling at 99 percent of the speed of light, your clock will register that seven hours pass while the spaceship's clocks show that only one hour has passed on board. This phenomenon is known as "time dilation," or "time dilatation," and it has been well-verified experimentally.

3) Mass is not an absolute measure, the same for all observers. For an object traveling at a fraction F of the speed of light, its mass will appear to be increased by a factor of $1/\sqrt{(1-F^2)}$ so far as an outside observer is concerned. If a spaceship passes you traveling at 99 percent of the speed of light, its mass will appear to have increased by a factor of seven over its original value. This phenomenon has also been well-verified experimentally.

4) Nothing can be accelerated to travel faster than light. In fact, to accelerate something to the speed of light would take an infinite amount of energy. This is actually a consequence of the previous point. *Note*: this does *not* say that an object cannot vanish from one place, and appear at another, in a time less than it would take light to travel between those locations. Hence this is not inconsistent with the quantum teleportation discussion of the previous section.

5) Length also is not an absolute measure, the same for all observers. If a spaceship passes you traveling at 99 percent of the speed of light, it will appear to be foreshortened to one-seventh of its original length. This phenomenon is known as "Lorentz contraction," or "Fitzgerald-Lorentz contraction."

6) The speed of light is the same for all observers, regardless of the speed of the light source or the speed of the observer. This is not so much a consequence of the special theory of relativity as one of the assumptions on which the theory is based.

The consequences of special relativity theory are worked out more simply if instead of dealing with space and time separately, calculations are performed in a merged entity we term "spacetime." This is also not a consequence of the theory, but rather a convenient framework in which to view it.

After it was proposed, the theory of relativity became the subject of much popular controversy. Detractors argued that the theory led to results that were preposterous and "obvious nonsense." That is not true, but certainly some of the consequences of relativity do not agree with "intuitive" common sense evolved by humans traveling at speeds very slow compared with the speed of light.

Let us consider just one example. Suppose that we have two spaceships, A and B, each traveling toward Earth (O), but coming from diametrically opposite directions. Also, suppose that each of them is moving at 4/5 of the speed of light according to their own measurements. "Common sense" would then insist that they are moving toward each other at 4/5+4/5=8/5 of light speed. Yet

one of our tenets for relativity theory is that you cannot
accelerate an object to the speed of light. But we seem
to have done just that. Surely, A will think that B is
approaching at 1.6 times light speed.

No. So far as A (or B) is concerned, we must use a
relativistic formula for combining velocities in order to calculate B's speed relative to A. According to that formula,
if O observes A and B approaching with speeds u and v,
then A (and B) will observe that they are approaching each
other at a speed $U=(u+v)/(1+uv/c^2)$, where c=the speed
of light. In this case, we find U=40/41 of the speed of
light.

Can U ever exceed c, for any permitted values of u
and v? That's the same as asking, if u and v are less
than 1, can $(u+v)/(1+uv)$ ever be greater than 1? It's
easy to prove that it cannot.

Now let us take the next step. Let us look at the
passage of time. Suppose that A sends a signal ("Hi")
and nine seconds later, according to his time frame, sends
the same message again. According to rule 2), above,
the time between one "Hi!" and the next, as measured
by us, will be increased by a factor 5/3. For us, 15
seconds have passed. And so far as B is concerned, since
B thinks that A is traveling at 40/41 of light-speed, an
interval $9/\sqrt{(1-40/41^2)}=41$ seconds have passed.

If you happen to be one of those people who read a
book from back to front, you may now be feeling confused. In discussing the expansion of the universe in
Chapter 4, we point out that signals from objects approaching us have higher frequencies, while signals from objects
receding from us have lower frequencies. But here we
seem to be saying the exact opposite of that: the time
between "Hi!" signals, which is equivalent to a frequency,
seems to be less for O and B than it is for A.

In fact, that is not the case. We have to allow for the
movement of A between transmission of successive
signals. When A sends the second "Hi," nine seconds
later than the first according to his measurements, he
has moved 15 seconds closer according to O. That is a
distance $15\times4/5=12$ light-seconds (a light-second, in
analogy to a light-year, is the distance light travels in

one second). Thus the travel time of the second "Hi"
is *decreased* by 12 seconds so far as O is concerned.
Hence the time between "Hi's" as measured by O is three
seconds. The signal frequency has increased.

The same is true for B. The time between transmission
of "Hi's" is 41 seconds as perceived by B, but in that
time the distance between A and B as measured by B
has decreased by 40 light-seconds. The time between
successive "Hi's" is therefore just one second for B. The
signal frequency so far as B is concerned has increased,
more than it did for O.

If the preceding few paragraphs seem difficult, don't
worry about them. My whole point is that the results
of relativity theory can be very counterintuitive when
your intuition was acquired in situations where everything
moves much less than the speed of light. The moral,
from a storyteller's point of view, is *be careful* when you
deal with objects or people moving close to light speed.
An otherwise good book, *The Sparrow* (Russell, 1996)
was ruined for me by a grotesque error in relativistic
time dilation effects. It could have been corrected with
a simple change of target star.

Just for the fun of it, let us ask what happens to our
signals between A, B, and O if we have a working
quantum teleportation device, able to send signals
instantaneously. What will the received signals have as
their frequencies? No one can give a definite answer to
this, but a likely answer is that quantum teleportation
is totally unaffected by relative velocities. If that's the
case, everyone sends and receives signals as though they
were all in close proximity and at rest relative to each
other. As a corollary, for quantum teleportation purposes
the universe lacks any spatial dimension and can be
treated as a single point.

2.6 General relativity. For the general theory of relativity, the main consequences to remember are:

1) The presence of matter (or of energy, which the
special theory asserts are two forms of one and the same
thing) causes space to curve. What we experience as
gravity is a direct measure of the curvature of space.

2) Objects move along the shortest possible path in curved space. Thus, a comet that falls in toward the Sun and then speeds out again following an elongated elliptical trajectory is traveling a minimum-distance path in curved space. In the same way, light that passes a massive gravitational object follows a path significantly bent from the straight line of normal geometry. Light that emanates from a massive object will be lengthened in wavelength as it travels "uphill" out of the gravity field. Note that this is not the "red shift" associated with the recession of distant galaxies, which will be discussed in Chapter 4.

3) If the concentration of matter is high, it is possible for spacetime itself to curve so much that a knowledge of some regions becomes denied to the rest of the universe. The interior of a black hole is just such a region. We are unaware of this in everyday life, simply because the concentrations of matter known to us are too low for the effects to occur.

4) Since matter curves space, the total amount of matter in the universe has an effect on its overall structure. This will become profoundly important in Chapter 4, when we consider the large-scale structure and eventual fate of the universe.

To truly space-faring civilizations, the effects of special and general relativity will be as much a part of their lives as sun, wind, and rain are to us.

2.7 Beyond the atom. Quantum theory and the special theory of relativity together provide the tool for analysis of subatomic processes. But we have not defined the subatomic world to which it applies.

Before the work of Rutherford and J.J. Thomson, the atom was considered a simple, indivisible object. Even after Rutherford's work, the only known subatomic particles were *electrons* and *protons* (the nucleus of an atom was regarded as a mixture of electrons and protons).

The situation changed in 1932, with the discovery of the *positron* (a positively charged electron) and the *neutron* (a particle similar in mass to the proton, but with no charge). At that point the atom came to be regarded as a cloud of electrons encircling a much smaller structure, the nucleus. The nucleus is made up of protons (equal in number to the electrons) and neutrons.

However, this ought to puzzle and worry us. Electrons and protons attract each other, because they are oppositely charged; but protons repel each other. So how is it possible for the nucleus, an assembly of protons and neutrons, to remain intact?

The answer is, through another force of nature, known as the *strong force*. The strong force is attractive, but it operates only over very short distances (much less than the size of an atom). It holds the nucleus together—most of the time. Sometimes another force, known as the *weak force*, causes a nucleus to emit an electron or a positron, and thereby become a different element. To round out the catalog of forces, we also have the familiar *electromagnetic force*, the one that governs attraction or repulsion of charged particles; and finally, we have the *gravitational force*, through which any two particles of matter, no matter how small or large, attract each other. The gravitational force ought really to be called the weak force, since it is many orders of magnitude less powerful than any of the others. It dominates the large-scale structure of the universe only because it applies over long distances, to every single bit of matter.

We have listed four fundamental forces. Are there others?

We know of no others, but that might only be an expression of our ignorance. From time to time, experiments hint at the existence of a "fifth force." Upon closer investigation, the evidence is explained some other way, and the four forces remain. However, it is quite legitimate in science fiction to hypothesize a fifth force, and to give it suitable properties. If you do this, however, be careful. The fifth force must be so subtle, or occur in such extreme circumstances, that we would not have stumbled over it already in our exploration of the universe. You can also, if you feel like it, suggest modifications to the existing four forces, again with suitable caution.

The attempt to explain the four known forces in a single framework, a Theory of Everything, assumes that no more forces will be discovered. This strikes me as a little presumptuous.

Returning to the discovery of fundamental particles, after the neutron came the *neutrino*, a particle with neither charge nor mass (usually; recently, some workers have discovered evidence suggesting a small mass for the neutrino). The existence of the neutrino had been postulated by Wolfgang Pauli in 1931, in order to retain the principle of the conservation of energy, but it was not actually discovered until 1953.

Then—too quickly for the theorists to feel comfortable about it—came the *muon* (1938), *pions* (predicted 1935, discovered 1947), the *antiproton* (1955), and a host of others, *etas* and *lambdas* and *sigmas* and *rhos* and *omegas*.

Quantum theory seemed to provide a theoretical framework suitable for all of these, but in 1960 the basic question—"Why are there so many animals in the 'nuclear zoo'?"—remained unanswered. In the early 1960s, Murray Gell-Mann and George Zweig independently proposed the existence of a new fundamental particle, the *quark*, from which all the heavy subatomic particles were made.

The quark is a peculiar object indeed. First, its charge is not a whole multiple of the electron or proton charge, but one-third or two-thirds of such a value. There are several varieties of quarks: the "up" and "down" quark, the "top" and "bottom" quark, and the "strange" and "charmed" quarks; each may have any of three "colors," red, green, or blue (the whimsical labels are no more than that; they lack physical significance). Taken together, the quarks provide the basis for a theory known as *quantum chromodynamics*, which is able to describe very accurately the forces that hold the atomic nucleus together.

A theory to explain the behavior of lighter particles (electrons, positrons, and photons) was developed earlier, mainly by Richard Feynman, Julian Schwinger, and Sinitiro Tomonaga. Freeman Dyson then proved the consistency and equivalence of the seemingly very different theories. The complete synthesis is known as *quantum electrodynamics*. Between them, quantum electrodynamics and quantum chromodynamics provide a full description of the subatomic world down to the scale at which we are able to measure.

However, the quark is a rather peculiar particle to employ as the basis for a theory. A proton consists of three quarks, two "up" and one "down"; a neutron is one "up" and two "down." Pions each contain only two quarks. An omega particle consists of three strange quarks. This is all based purely on theory, because curiously enough, no one has ever seen a quark. Theory suggests that we never will. The quark exists *only in combination with other quarks*. If you try to break a quark free, by hitting a proton or a neutron with high-energy electrons or a beam of radiation, at first nothing appears. However, if you keep increasing the energy of the interaction, something finally does happen. New particles appear—not the quarks, but more protons, pions, and neutrons. Energy and mass are interchangeable; apply enough energy, and particles appear. The quark, however, keeps its privacy.

I have often thought that a good bumper sticker for a particle physicist would be "Free the Quarks!"

The reluctance of the free quark to put in an appearance makes it very difficult for us to explore its own composition. But we ask the question: What, if anything, is smaller than the quark?

Although recent experiments suggest that the quark does have a structure, no one today knows what it is. We are offshore of the physics mainland, and are allowed to speculate in fictional terms as freely as we choose.

Or almost. There are two other outposts that we need to be aware of in the world of the ultra-small. The proton and the neutron have a radius of about 0.8×10^{-15} meters. If we go to distances far smaller than that, we reach the realm of the *superstring*.

A superstring is a loop of something not completely defined (energy? force?) that oscillates in a space of ten dimensions. The string vibrations give rise to all the known particles and forces. Each string is about 10^{-35} meters long. We live in a space of four dimensions (three

space and one time), and the extra six dimensions of
superstring theory are "rolled up" in such a way as to
be unobservable. In practice, of course, a superstring has
never been observed. The necessary mathematics to
describe what goes on is also profoundly difficult.

Why is the concept useful? Mainly, because superstring
theory includes gravity in a natural way, which quantum
electrodynamics and quantum chromodynamics do not.
In fact, superstring theory not only allows gravity to be
included, it requires it. We might be closing in on the
"Theory of Everything" already mentioned, explaining the
four known fundamental interactions of matter in a single
set of equations.

There is a large literature on superstrings. If the concept
continues to prove useful, we will surely find ways to
make the mathematics more accessible. Remember, the
calculus needed to develop Isaac Newton's theories was
considered impossibly difficult in the seventeenth century.
Meanwhile, the science fiction writer can be more
comfortable with superstrings than many practicing
scientists.

On the same small scale as the superstring we have
something known as the *Planck length*. This is the place
where vacuum energy fluctuations, inevitably predicted
by quantum theory, radically affect the nature of space.
Rather than a smooth, continuous emptiness, the
vacuum must now be perceived as a boiling chaos of
minute singularities. Tiny black holes constantly form
and dissolve, and space has a foam-like character where
even the concept of distance may not have meaning.
(We have mentioned black holes but not really discussed them, though surely there is no reader who has
not heard of them. They are so important a part of
the science fiction writer's arsenal that they deserve a
whole section to themselves. They can be found in
Chapter 3.)

So far as science is concerned, the universe at the scale
of the Planck length is true *terra incognita*, not to be
found on any map. I know of no one who has explored
its story potential. You, as storyteller, are free to roam
as you choose.

2.8 Strange physics: superconductivity. I was not sure where this ought to be in the book. It is a phenomenon which depends on quantum level effects, but its results show up in the macroscopic world of everyday events. The one thing that I was sure of is that this is too fascinating a subject to leave out, something that came as an absolute and total surprise to scientists when it was discovered, and remained a theoretical mystery for forty years thereafter. If superconductivity is not a fertile subject for writers, nothing is.

Superconductivity was first observed in materials at extremely low temperatures, so that is the logical place to begin.

Temperature, and particularly low temperature, is in historical terms relatively new. Ten thousand years ago, people already knew how to make things hot. It was easy. You put a fire underneath them. But as recently as two hundred years ago, it was difficult to make things cold. There was no "cold generator" that corresponded to fire as a heat generator. Low temperatures were something that came naturally, they were not man-made.

The Greeks and Romans knew that there were ways of lowering the temperature of materials, although they did not use that word, by such things as the mixture of salt and ice. But they had no way of seeking progressively lower temperatures. That had to wait for the early part of the nineteenth century, when Humphrey Davy and others found that you could liquefy many gases merely by compressing them. The resulting liquid will be warm, because turning gas to liquid gives off the gas's so-called "latent heat of liquefaction." If you now allow this liquid to reach a thermal balance with its surroundings, and then reduce the pressure on it, the liquid boils; and in so doing, it drains heat from its surroundings—including itself. The same result can be obtained if you take a liquid at atmospheric pressure, and put it into a partial vacuum. Some of the liquid boils, and what's left is colder. This technique, of "boiling under reduced pressure," was a practical and systematic way of pursuing lower temperatures. It first seems to have been used by a Scotsman, William Cullen, who cooled ethyl ether this way in 1748, but it took another three-quarters of a century before the method was applied to science (and to commerce; the first refrigerator was patented by Jacob Perkins in 1834).

Another way to cool was found by James Prescott Joule and William Thomson (later Lord Kelvin) in 1852. Named the Joule-Thomson effect, or the Joule-Kelvin effect, it relies on the fact that a gas escaping from a valve into a chamber of lower pressure will, under the right conditions, suffer a reduction in temperature. If the gas entering the valve is first passed in a tube through that lower-temperature region, we have a cycle that will move the chamber to lower and lower temperatures.

Through the nineteenth century the Joule-Thomson effect and boiling under reduced pressure permitted the exploration of lower and lower temperatures. The natural question was, how low could you go?

A few centuries ago, there seemed to be no answer to that question. There seemed no limit to how cold something could get, just as today there is no practical limit to how hot something can become.

The problem of reaching low temperatures was clarified when scientists finally realized, after huge intellectual efforts, that heat is nothing more than motion at the atomic and molecular scale. "Absolute zero" could then be identified as no motion, the temperature of an object when you "took out all the heat." (Purists will object to this statement since even at absolute zero, quantum theory tells us that an object still has a zero point energy; the thermodynamic definition of absolute zero is done in terms of reversible isothermal processes.)

Absolute zero, it turns out, is reached at a temperature of −273.16 degrees Celsius. Temperatures measured with respect to this value are all positive, and are said to be in *Kelvins* (written K). One Kelvin is the same *size* as one degree Celsius, but it is measured with respect to a reference point of absolute zero, rather than to the Celsius zero value of the freezing point of water. We will use the two scales interchangeably, whichever is the more convenient at the time.

Is it obvious that this absolute zero temperature must be the same for all materials? Suppose that you had two materials which reached their zero heat state at different temperatures. Put them in contact with each other. Then thermodynamics requires that heat should flow from the higher temperature body to the other one, until they both reach the same temperature. Since there is by assumption no heat in either material (each is at its own absolute zero), no heat can flow; and when no heat flows between two bodies in contact, they must be at the same temperature. Thus absolute zero is the same temperature for every material.

Even before an absolute zero point of temperature was identified, people were trying to get down as low in temperature as they could, and also to liquefy gases. Sulfur dioxide (boiling point $-10°C$) was the first to go, when Monge and Clouet liquefied it in 1799 by cooling in a mixture of ice and salt. De Morveau produced liquid ammonia (boiling point $-33°C$) in 1799 using the same method, and in 1805 Northmore claimed to have produced liquid chlorine (boiling point $-35°C$) by simple compression.

In 1834, Thilorier produced carbon dioxide snow (dry ice, melting point $-78.5°C$) for the first time using gas expansion. Soon after that, Michael Faraday, who had earlier (1823) liquefied chlorine, employed a carbon dioxide and ether mixture to reach the record low temperature of -110 degrees Celsius (163 K). He was able to liquefy many gases, but not hydrogen, oxygen, or nitrogen.

In 1877, Louis Cailletet used gas compression to several hundred atmospheres, followed by expansion through a jet, to produce liquid mists of methane (boiling point $-164°C$), carbon monoxide (boiling point $-192°C$), and oxygen (boiling point $-183°C$). He did not, however, manage to collect a volume of liquid from any of these substances.

Liquid oxygen was finally produced in quantity in 1883, by Wroblewski and Olszewski, who reached the lowest temperature to date ($-136°C$). Two years later they were able to go as low as $-152°C$, and liquefied both nitrogen and carbon monoxide. In that same year, Olszewski reached a temperature of −225°C (48 K), which remained a record for many years. He was able to produce a small amount of liquid hydrogen for the first time. In 1886, Joseph Dewar invented the Dewar flask (which we think of today as the thermos bottle) that allowed cold, liquefied materials to be stored for substantial periods of time at atmospheric pressure. In 1898, Dewar liquefied hydrogen in quantity and reached a temperature of 20 K. At that point, all known gases had been liquefied.

I have gone a little heavy on the history here, to make the point that most scientific progress is not the huge intellectual leap favored in bad movies. It is more often careful experiments and the slow accretion of facts, until finally one theory can be produced which encompasses all that is known. If a story is to be plausible and involves a major scientific development, then some (invented) history that preceded the development adds a feeling of reality.

However, we have one missing fact in the story so far. What about helium, which has not been mentioned?

In the 1890s, helium was still a near-unknown quantity. The gas had been observed in the spectrum of the Sun by Janssen and Lockyer, in 1868, but it had not been found on earth until the early 1890s. Its properties were not known. It is only with hindsight that we can find good reasons why the gas, when available, proved unusually hard to liquefy.

The periodic table had already been formulated by Dmitri Mendeleyev, in about 1870. Forty years later, Henry Moseley showed that the table could be written in terms of an element's *atomic number*, which corresponded to the number of protons in the nucleus of that element.

As other gases were liquefied, a pattern emerged. TABLE 2.1 (p. 57) shows the temperatures where a number of gases change from the gaseous to the liquid state, under normal atmospheric pressure, together with their atomic numbers and molecular weights.

What happens when we plot the boiling point of an element against its atomic number in the periodic table? For gases, there are clearly two different groups. Radon, xenon, krypton, argon, and neon remain gases to much lower temperatures than other materials of similar atomic number. This is even more noticeable if we add a number of other common gases, such as ammonia, acetylene, carbon dioxide, methane, and sulfur dioxide, and look at the variation of their boiling points with their molecular weights. They all boil at much higher temperatures.

Now, radon, xenon, krypton, and the others of the low-boiling-point group are all inert gases, often known as noble gases, that do not readily participate in any chemical reactions. TABLE 2.1 (p. 57) also shows that the inert gases of lower atomic number and molecular weight liquefy at lower temperatures. Helium, the second lightest element, is the final member of the inert gas group, and the one with the lowest atomic number. Helium should therefore have an unusually low boiling point.

It does. All through the late 1890s and early 1900s, attempts to liquefy it failed.

When the Dutch scientist Kamerlingh Onnes finally succeeded, in 1908, the reason for other people's failure became clear. Helium remains liquid until -268.9 Celsius—16 degrees lower than liquid hydrogen, and only 4.2 degrees above absolute zero. As for solid helium, not even Onnes' most strenuous efforts could produce it. When he boiled helium under reduced pressure, the liquid helium went to a new form—but it was a new and strange liquid phase, now known as Helium II, that exists only below 2.2 K. It turns out that the solid phase of helium does not exist at atmospheric pressure, or at any pressure less than 25 atmospheres. It was first produced in 1926, by P.H. Keeson.

The liquefaction of helium looked like the end of the story; it was in fact the beginning.

2.9 Super properties. Having produced liquid helium, Kamerlingh Onnes set about determining its properties. History does not record what he expected to find, but it is fair to guess that he was amazed.

Science might be defined as assuming something you don't know using what you know, and then measuring to see if it is true or not. The biggest scientific advances often occur when what you measure does not agree with what you predict. What Kamerlingh Onnes measured for liquid helium, and particularly for Helium II, was so bizarre that he must have wondered at first what was wrong with his measuring equipment.

One of the things that he measured was viscosity. Viscosity is the gooeyness of a substance, though there are more scientific definitions. We usually think of viscosity as applying to something like oil or molasses, but non-gooey substances like water and alcohol have well-defined viscosities.

Onnes tried to determine a value of viscosity for Helium II down around 1 K. He failed. It was too small to measure. As the temperature goes below 2 K, the viscosity of Helium II goes rapidly towards zero. It will flow with no measurable resistance through narrow capillaries and closely-packed powders. Above 2.2 K, the other form of liquid helium, known as Helium I, does have a measurable viscosity, low but highly temperature-dependent.

Helium II also conducts heat amazingly well. At about 1.9 K, where its conductivity is close to a maximum, this form of liquid helium conducts heat about eight hundred times as well as copper at room temperature— and copper is usually considered an excellent conductor. Helium II is in fact by far the best known conductor of heat.

More disturbing, perhaps, from the experimenter's point of view is Helium II's odd reluctance to be confined. In an open vessel, the liquid creeps in the form of a thin film up the sides of the container, slides out over the rim, and runs down to the lowest available level. This phenomenon can be readily explained, in terms of the very high surface tension of Helium II; but it remains a striking effect to observe.

Liquid helium is not the end of the low-temperature story, and the quest for absolute zero is an active and fascinating field that continues today. New methods of extracting energy from test substances are still being developed, with the most effective ones employing a technique known as adiabatic demagnetization. Invented independently in 1926 by a German, Debye, and an

American, Giauque, it was first used by Giauque and MacDougall in 1933, to reach a temperature of 0.25 K. A more advanced version of the same method was applied to nuclear adiabatic demagnetization in 1956 by Simon and Kurti, and they achieved a temperature within a hundred thousandth of a degree of absolute zero. With the use of this method, temperatures as low as a few billionths of a degree have been attained.

However, the pursuit of absolute zero is not our main objective, and to pursue it further would take us too far afield. We are interested in another effect that Kamerlingh Onnes found in 1911, when he examined the electrical properties of selected materials immersed in a bath of liquid helium. He discovered that certain pure metals exhibited what is known today as *superconductivity*.

Below a few Kelvins, the resistance to the passage of an electrical current in these metals drops suddenly to a level too small to measure. Currents that are started in wire loops under these conditions continue to flow, apparently forever, with no sign of dissipation of energy. For pure materials, the cutoff temperature between normal conducting and superconducting is quite sharp, occurring within a couple of hundredths of a degree. Superconductivity today is a familiar phenomenon. At the time when it was discovered, it was an absolutely astonishing finding—a physical impossibility, less plausible than anti-gravity. Frictional forces must slow all motion, including the motion represented by the flow of an electrical current. Such a current could not therefore keep running, year after year, without dissipation. That seemed like a fundamental law of nature.

Of course, there is no such thing as a law of nature. There is only the Universe, going about its business, while humans scurry around trying to put everything into neat little intellectual boxes. It is amazing that the tidying-up process called physics works as well as it does, and perhaps even more astonishing that mathematics seems important in every box. But the boxes have no reality or permanence; a "law of nature" is useful until we discover cases where it doesn't apply.

In 1911, the general theories that could explain superconductivity were still decades in the future. The full explanation did not arrive until 1957, forty-six years after the initial discovery.

To understand superconductivity, and to explain its seeming impossibility, it is necessary to look at the nature of electrical flow itself.

2.10 Meanwhile, electricity. While techniques were being developed to reach lower and lower temperatures, the new field of electricity and magnetism was being explored in parallel—sometimes by the same experimenters. Just three years before the Scotsman, William Cullen, found how to cool ethyl ether by boiling it under reduced pressure, von Kleist of Pomerania and van Musschenbroek in Holland independently discovered a way to store electricity. Van Musschenbroek did his work at the University of Leyden—the same university where, 166 years later, Kamerlingh Onnes would discover superconductivity. The Leyden Jar, as the storage vessel soon became known, was an early form of electrical capacitor. It allowed the flow of current through a wire to take place under controlled and repeatable circumstances.

Just what it was that constituted the current through that wire would remain a mystery for another century and a half. But it was already apparent to Ben Franklin by 1750 that *something* material was flowing. The most important experiments took place three-quarters of a century later. In 1820, just three years before Michael Faraday liquefied chlorine, the Danish scientist Hans Christian Oersted and then the Frenchman André Marie Ampère found that there was a relation between electricity and magnetism—a flowing current would make a magnet move. In the early 1830s, Faraday then showed that the relationship was a reciprocal one, by producing an electric current from a moving magnet. However, from our point of view an even more significant result had been established a few years before, when in 1827 the German scientist Georg Simon Ohm discovered Ohm's Law: that the current in a wire is given by the ratio of the voltage between the ends of the wire, divided by the wire's resistance.

This result seemed too simple to be true. When Ohm announced it, no one believed him. He was discredited, resigned his position as a professor at Cologne University, and lived in poverty and obscurity for several years. Finally, he was vindicated, recognized, and fourteen years later began to receive awards and medals for his work.

Ohm's Law is important to us because the resistance of a substance does not depend on the particular values of the voltage or the current. Thus it becomes a simple matter to study the dependence of resistance on temperature. It turns out that the resistance of a conducting material is roughly proportional to its absolute temperature. Just as important, materials vary enormously in their conducting power. For instance, copper allows electricity to pass through it 10^{20} times as well as quartz or rubber. The obvious question is, why? What makes a good conductor, and what makes a good insulator? And why should a conductor pass electricity more easily at lower temperatures?

The answers to these questions were developed little by little through the rest of the nineteenth century. First, heat was discovered to be no more than molecular and atomic motion. Thus changes of electrical resistance had somehow to be related to those same motions.

Second, in the 1860s, Maxwell, the greatest physicist of the century, developed Faraday and Ampère's experimental results into a consistent and complete mathematical theory of electricity and magnetism, finally embodied in four famous differential equations. All observed phenomena of electricity and magnetism must fit into the framework of that theory.

Third, scientists began to realize that metals, and many other materials that conduct electricity well, have a regular structure at the molecular level. The atoms and molecules of these substances are arranged in a regular three-dimensional grid pattern, termed a lattice, and held in position by interatomic electrical forces.

Finally, in 1897, J.J. Thomson found the elusive carrier of the electrical current. He originally termed it the "corpuscle," but it soon found its present name, the electron. All electrical currents are carried by electrons.

Again, lots of history before we have the tools in hand to understand the flow of electricity through conductors—but not yet, as we shall see, to explain superconductivity.

Electricity is caused by the movement of electrons. Thus a good conductor must have plenty of electrons readily able to move, which are termed free electrons. An insulator has few or no free electrons, and the electrons in such materials are all bound to atoms.

If the atoms of a material maintain exact, regularly spaced positions, it is very easy for free electrons to move past them, and hence for current to flow. In fact, electrons are not interfered with at all if the atoms in the material stand in a perfectly regular array. However, if the atoms in the lattice can move randomly, or if there are imperfections in the lattice, the electrons are then impeded in their progress, and the resistance of the material increases.

This is exactly what happens when the temperature goes up. Recalling that heat is random motion, we expect that atoms in hot materials will jiggle about on their lattice sites with the energy provided by increased heat. The higher the temperature, the greater the movement, and the greater the obstacle to free electrons. Therefore the resistance of conducting materials increases with increasing temperature.

This was all well-known by the 1930s. Electrical conduction could be calculated very well by the new quantum theory, thanks largely to the efforts of Arnold Sommerfeld, Felix Bloch, Rudolf Peierls, and others. However, those same theories predicted a steady decline of electrical resistance as the temperature went towards absolute zero. Nothing predicted, or could explain, the precipitous drop to zero resistance that was encountered in some materials at their critical temperature. Superconductivity remained a mystery for another quarter of a century. To provide its explanation, it is necessary to delve a little further into quantum theory itself.

2.11 Superconductivity and statistics. Until late 1986, superconductivity was a phenomenon never encountered at temperatures above 23 K, and usually at just a couple of degrees Kelvin. Even 23 K is below the boiling point of everything except hydrogen (20 K) and helium (4.2 K). Most superconductors become so only at far lower temperatures (see TABLE 2.2). Working with them is thus a tiresome business, since such low temperatures are expensive to achieve and hard to maintain. Let us term superconductivity below 20 K "classical superconductivity," and for the moment confine our attention to it. TABLE 2.2 shows the temperature at which selected materials become superconducting when no magnetic field is present.

Note that all these temperatures are below the temperature of liquid hydrogen (20 K), which means that superconductivity cannot be induced by bathing the metal sample in a liquid hydrogen bath, although such an environment is today readily produced. For many years, the search was for a material that would sustain superconductivity above 20 K.

For another fifteen years after the 1911 discovery of superconductivity, there seemed little hope of explaining it. However, in the mid-1920s a new tool, quantum theory, encouraged physicists to believe that they at last had a theoretical framework that would explain all phenomena of the subatomic world. In the late 1920s and 1930s, hundreds of previously-intractable problems yielded to a quantum mechanical approach. And the importance of a new type of statistical behavior became clear.

On the atomic and nuclear scale, particles and systems of particles can be placed into two well-defined and separate groups. Electrons, protons, neutrons, positrons, muons, and neutrinos all satisfy what is known as Fermi-Dirac statistics, and they are collectively known as fermions. For our purposes, the most important point about such particles is that their behavior is subject to the Pauli Exclusion Principle, which states that no two identical particles obeying Fermi-Dirac statistics can have the same values for all physical variables (so, for example, two electrons in an atom cannot have the same spin, the same angular momentum, and the same energy level). The Pauli Exclusion Principle imposes very strong constraints on the motion and energy levels of identical fermions, within atoms and molecules, or moving in an atomic lattice.

The other kind of statistics is known as Bose-Einstein statistics, and it governs the behavior of photons, alpha particles (i.e. helium nuclei), and mesons (pions and some other subnuclear particles). These are all termed bosons. The Pauli Exclusion Principle does not apply to systems satisfying Bose-Einstein statistics, so bosons are permitted to have the same values of all physical variables; in fact, since they seek to occupy the lowest available energy level, they will group around the same energy.

In human terms, fermions are loners, each with its own unique state; bosons love a crowd, and they all tend to jam into the same state.

Single electrons are, as stated, fermions. At normal temperatures, which are all well above a few Kelvins, electrons in a metal are thus distributed over a range of energies and momenta, as required by the Pauli Exclusion Principle.

In 1950, H. Fröhlich suggested a strange possibility: that the fundamental mechanism responsible for super-conductivity was somehow the interaction of free electrons with the atomic lattice. This sounds at first sight highly improbable, since it is exactly this lattice that is responsible for the resistance of metals at normal temperatures. However, Fröhlich had theoretical reasons for his suggestion, and in that same year, 1950, there was experimental evidence—unknown to Fröhlich—that also suggested the same thing: superconductivity is caused by electron-lattice interactions.

This does not, of course, explain superconductivity. The question is, what does the lattice do? What can it possibly do, that would give rise to superconducting materials? Somehow the lattice must affect the free electrons in a fundamental way, but in a way that is able to produce an effect only at low temperatures.

The answer was provided by John Bardeen, Leon

Cooper, and Robert Schrieffer, in 1957 (they got the physics Nobel prize for this work in 1972). They showed that the atomic lattice causes free electrons to pair off. Instead of single electrons, moving independently of each other, the lattice encourages the formation of electron couplets, which can then each be treated as a unit. The coupling force is tiny, and if there is appreciable thermal energy available it is enough to break the bonds between the electron pairs. Thus any effect of the pairing should be visible only at very low temperatures. The role of the lattice in this pairing is absolutely fundamental, yet at the same time the lattice does not participate in the pairing—it is more as if the lattice is a catalyst, which permits the electron pairing to occur but is not itself affected by that pairing.

The pairing does not mean that the two electrons are close together in space. It is a pairing of angular momentum, in such a way that the total angular momentum of a pair is zero. The two partners may be widely separated in space, with many other electrons between them; but, like husbands and wives at a crowded party, paired electrons remain paired even when they are not close together.

Now for the fundamental point implied by the work of Cooper, Bardeen, and Schrieffer. Once two electrons are paired, that pair behaves like a boson, not a fermion. Any number of these electron pairs can be in the same low-energy state. More than that, when a current is flowing (so all the electron pairs are moving) it takes more energy to stop the flow than to continue it. To stop the flow, some boson (electron pair) will have to move to a different energy level; and as we already remarked, bosons like to be in the same state.

To draw the chain of reasoning again: superconductivity is a direct result of the boson nature of electron pairs; electron pairs are the direct result of the mediating effect of the atomic lattice; and the energy holding the pairs together is very small, so that they exist only at very low temperatures, when no heat energy is around to break up the pairing.

2.12 High-temperature superconductors. We now have a very tidy explanation of classical superconductivity, one that suggests we will never find anything that behaves as a superconductor at more than a few degrees above absolute zero. Thus the discovery of materials that turn into superconductors at much higher temperatures is almost an embarrassment. Let's look at them and see what is going on.

The search for high-temperature superconductors began as soon as superconductivity itself was discovered. Since there was no good theory before the 1950s to explain the phenomenon, there was also no reason to assume that a material could not be found that exhibited superconductivity at room temperature, or even above it. That, however, was not the near-term goal. The main hope of researchers in the field was more modest, to find a material with superconductivity well above the temperature of liquid hydrogen. Scientists would certainly have loved to find something better yet, perhaps a material that remained superconducting above the temperature of liquid nitrogen (77 K). That would have allowed superconductors to be readily used in many applications, from electromagnets to power transmission. But as recently as December 1986, that looked like an impossible dream.

The first signs of the breakthrough had come early that year. In January 1986, Alex Müller and Georg Bednorz, at the IBM Research Division in Zurich, Switzerland, produced superconductivity in a ceramic sample containing barium, copper, oxygen, and lanthanum (one of the rare-earth elements). The temperature was 11 K, which was not earth-shaking, but much higher than anyone might have expected. Müller and Bednorz knew they were on to something good. They produced new ceramic samples, and little by little worked the temperature for the onset of superconductivity up to 30 K. The old record, established in 1973, had been 23 K. By November, Paul Chu and colleagues at the University of Houston, and Tanaka and Kitazawa at the University of Tokyo had repeated the experiments, and also found the material superconducting at 30 K.

Once those results were announced, every experimental team engaged in superconductor research jumped onto the problem. In December, Robert Cava, Bruce van Dover, and Bertram Batlogg at Bell Labs had produced superconductivity in a strontium-lanthanum-copper-oxide combination at 36 K. Also in December, 1986, Chu and colleagues had positive results over 50 K.

In January 1987, there was another astonishing breakthrough. Chu and his fellow workers substituted yttrium, a metal with many rare-earth properties, for lanthanum in the ceramic pellets they were making. The resulting samples went superconducting at 90 K. The researchers could hardly believe their result, but within a few days they had pushed up to 93 K, and had a repeatable, replicable procedure. Research groups in Tokyo and in Beijing also reported results above 90 K in February.

Recall that liquid nitrogen boils at 77 K. For the first time, superconductors had passed the "nitrogen barrier." In a bath of that liquid, a ceramic wire using yttrium, barium, copper, and oxygen was superconducting.

The end of the road has still not been reached. There have been hints of superconductive behavior at 234 K. This is only $-40°C$, just a few degrees below the temperature at which ammonia boils.

Fascinating, and the natural question is, can room-temperature superconductors, the Holy Grail of this field, ever be produced?

Unfortunately, the question cannot be answered. There is no accepted model to explain what is going on, and it would not be unfair to say that at the moment experiment is still ahead of theory.

The Bardeen, Cooper, and Schrieffer (BCS) theory of superconductivity leads to a very weak binding force between electron pairs. Thus according to this theory the phenomenon ought not to occur at 90 K, still less at 240 K. At the same time, the theory tells us that *any* superconductivity, high-temperature or otherwise, is almost certainly the result of free electrons forming into pairs, and then behaving as bosons. In classical superconductivity, at just a few degrees above absolute zero, the mediating influence that operates to form electron pairs can be shown to be the atomic lattice itself. That result, in quantitative form, comes from the Cooper, Bardeen, and Schrieffer approach. The natural question to ask is, What other factor could work to produce electron pairs? To be useful, it must produce strong bonding of electron pairs, otherwise they would be dissociated by the plentiful thermal energy at higher temperatures. And any electron pairs so produced must be free to move, in order to carry the electric current.

Again, we are asking questions that take us beyond the frontiers of today's science. Any writer has ample scope for speculation.

2.13 Making it work. Does this mean that we now have useful, workhorse superconductors above the temperature of liquid nitrogen, ready for industrial applications? It looks like it. But there are complications.

Soon after Kamerlingh Onnes discovered superconductivity, he also discovered (in 1913) that superconductivity was destroyed when he sent a large current through the material. This is a consequence of the effect that Oersted and Ampère had noticed in 1820; namely, that an electric current creates a magnetic field. The temperature at which a material goes superconducting is lowered when it is placed in a magnetic field. That is why the stipulation was made in TABLE 2.2 (p. 57) that those temperatures apply only when no magnetic field is present. A large current creates its own magnetic field, so it may itself destroy the superconducting property.

For a superconductor to be useful in power transmission, it must remain superconducting even though the current through it is large. Thus we want the critical temperature to be insensitive to the current through the sample. One concern was that the new high-temperature superconductors might perform poorly here, and the first samples made were in fact highly affected by imposed magnetic fields. However, some of the new superconductors have been found to remain superconducting at current up to 1,000 amperes per square millimeter, and this is more than adequate for power transmission.

A second concern is a practical one: Can the new materials be worked with, to make wires and coils that are not too brittle or too variable in quality? Again the answers are positive. The ceramics can be formed into loops and wires, and they are not unduly brittle or fickle in behavior.

The only thing left is to learn where the new capability of high-temperature superconductors will be most useful. Some applications are already clear.

First, we will see smaller and faster computers, where the problem of carrying off heat caused by dissipation of energy from electrical currents in small components is a big problem. This application will exist, even if the highest temperature superconductors cannot tolerate high current densities.

Second, as Faraday discovered, any tightly-wound coil of wire with a current running through it becomes an electromagnet. Superconducting coils can produce very powerful magnets of this type, ones which will keep their magnetic properties without using any energy or needing any cooling. Today's electromagnets that operate at room temperature are limited in their strength, because very large currents through the coils also produce large heating effects.

Third, superconductors have another important property that we have not so far mentioned, namely, they do not allow a magnetic field to be formed within them. In the language of electromagnetic theory, they are *perfectly diamagnetic*. This is known as the Meissner Effect, and it was discovered in 1933. It could have easily been found in 1913, but it was considered so unlikely a possibility that no one did the experiment to test superconductor diamagnetism for another twenty years.

As a consequence of the Meissner Effect, a superconductor that is near a magnet will form an electric current layer on its surface. That layer is such that the superconductor is then strongly *repelled* by the magnetic field, rather than being attracted to it. This permits a technique known as *magnetic levitation* to be used to lift and support heavy objects. Magnets, suspended above a line of superconductors, will remain there without needing any energy to hold them up. Friction-free support systems are the result, and they should be useful in everything from transportation to factory assembly lines. For many years, people have talked of super-speed trains, suspended by magnetic fields and running at a fraction of the operating costs of today's locomotives. When superconductors could operate only when cooled to liquid hydrogen temperatures and below, such transportation ideas were hopelessly expensive. With high-temperature superconductors, they become economically attractive.

And of course, there is the transmission of electrical power. Today's transmission grids are full of transformers that boost the electrical signal to hundreds of thousands of volts for sending the power through the lines, and then bring that high-voltage signal back to a hundred volts or so for household use. However, the only reason for doing this is to minimize energy losses. Line heating is less when electrical power is transmitted at low current and high voltage, so the higher the voltage, the better. With superconductors, however, there are no heat dissipation losses at all. Today's elaborate system of transformers will be unnecessary. The implications of this are enormous: the possible replacement of the entire electrical transmission systems of the world by a less expensive alternative, both to build and to operate.

However, before anyone embarks on such an effort, they will want to be sure that the technology has gone as far as it is likely to go. It would be crazy to start building a power-line system based on the assumption that the superconductors need to be cooled to liquid nitrogen or liquid ammonia temperatures, if next year sees the discovery of a material that remains super-conducting at room temperature and beyond.

Super-computers, heavy lift systems, magnetically cushioned super-trains, cheap electrical power transmission; these are the obvious prospects. Are there other important uses that have not yet been documented?

Almost certainly, there are. We simply have to think of them; and then, before scientists prove that our ideas are impossible, it would be nice to write and publish stories about them. It will not be enough, by the way, to simply refer to room-temperature superconductors. That was done long ago, by me among others ("A Certain Place In History," *Galaxy*, 1977).

TABLE 2.1
The boiling points of gases.

Gas	Boiling point (C)	(K)	Atomic number	Molecular weight
Radon	−61.8	211.4	86	222
Xenon	−107.1	166.1	54	131
Krypton	−152.3	120.9	36	84
Argon	−185.7	87.5	18	40
Chlorine	−34.6	238.6	17	71
Neon	−246.1	27.1	10	20
Fluorine	−188.1	85.1	9	38
Oxygen	−183.0	90.2	8	32
Nitrogen	−195.8	77.3	7	28
Hydrogen	−252.8	20.4	1	2

TABLE 2.2.
Temperatures at which materials become superconducting (in the absence of a magnetic field).

Material	Temperature (K)
Titanium	0.39
Zinc	0.93
Uranium	1.10
Aluminum	1.20
Tin	3.74
Mercury	4.16
Lead	7.22
Niobium	8.90
Technetium	11.20

CHAPTER 3
Physics in the Large

3.1 Stars. Everything between atoms and stars, roughly speaking, belongs to chemistry. Although you and I are certainly subject to the laws of physics, we are chemical objects. Our metabolism and structure are controlled by the laws of chemistry. The same is largely true of planets. The shape of the Earth is defined by gravity, but most of the activities within it, or on its surface, or in its atmosphere, are decided by the laws of chemistry.

This is not true of stars. To understand how a star like the Sun can shine for billions of years, you need physics.

The modern view of stars, as giant globes of hot gas, began in 1609, when Galileo Galilei turned his home-made telescope upwards. Rather than a perfect sphere whose nature defied explanation, Galileo found that the Sun was a rotating object with lots of surface detail like sunspots and solar flares.

Over the next couple of hundred years, the size and the temperature of the sun were determined. It is a ball of gas, about a million miles across, with a surface at 6,000 degrees Celsius. What was not understood at all, even a hundred years ago, was the way that the sun stays hot.

Before 1800, that was not a worry. The universe was believed to be only a few thousand years old (Archbishop Ussher of Armagh, working through the genealogy of the

Bible, in 1654 announced that the time of creation was 4,004 B.C., on October 26th. No messing about with uncertainty for him.)

In the eighteenth century, the scriptural time-scale prevented anyone worrying much about the age of the Sun. If it had started out very hot in 4000 B.C., it hadn't had time to cool down yet. If it were made entirely of burning coal, it would have lasted long enough. A chemical explanation was adequate.

Around 1800, the geologists started to ruin things. In particular, James Hutton proposed his theory of geological *uniformitarianism* (Hutton, 1795).

Uniformitarianism, in spite of its ugly name, is a beautiful and simple idea. According to Hutton, the processes that built the world in the past are exactly those at work today: the uplift of mountains, the tides, the weathering effects of rain and air and water flow, these shape the surface of the Earth. This is in sharp distinction to the idea that the world was created just as it is now, except for occasional great catastrophic changes like the Biblical Flood.

The great virtue of Hutton's theory is that it removes the need for assumptions. Anything that shaped the past can be assessed by looking at its effectiveness today.

The great disadvantage of the theory, from the point of view of anyone pondering what keeps the Sun hot, is the amount of time it takes for all this to happen. We can no longer accept a universe only a few thousand years old. Mountain ranges could not form, seabeds be raised, chalk deposits laid down, and solid rocks erode to powder, in so short a time. Millions of years, at a minimum, are needed.

A Sun made of coal will not do. Nothing chemical will do. In the 1850s, Hermann von Helmholtz and Lord Kelvin finally proposed a solution, drawn from physics, that could give geology more time. They suggested that the source of the Sun's heat was *gravitational contraction*. If the material of the Sun were slowly falling inward on itself, that would release energy. The amount of energy produced by the Sun's contraction could be precisely calculated.

Unfortunately, it was still not enough. While Lord Kelvin was proposing an age for the Sun of 20 million years, the ungrateful geologists, and still more so the biologists, were asking considerably more. Charles Darwin's *Origin of Species* came out in 1859, and evolution seemed to need much longer than mere tens of millions of years to do its work. The biologists wanted hundreds of millions at a minimum; they preferred a few billion.

No one could give it to them during the whole of the nineteenth century. Lord Kelvin, who no matter what he did could not come up with any age for the Sun greater than 100 million years and was in favor of a number far less, became an archenemy of the evolutionists. An "odious spectre" is what Darwin called him. But no one could refute his physical arguments. A scientific revolution was needed before an explanation was available for a multibillion-year age of the Sun.

That revolution began, as we saw, in the 1890s. The atom, previously thought indivisible, had an interior structure and could be broken into smaller pieces. By the 1920s it was realized that lightweight atoms could also *combine*, to form heavier atoms. In particular, four atoms of hydrogen could fuse together to form one atom of helium; and if that happened, huge amounts of energy could be produced.

Perhaps the first person to realize that nuclear fusion was the key to what makes the sun go on shining was Eddington. Certainly he was one of the first persons to develop the idea systematically, and equally certainly he believed that he was the first to think of it. There is a story of Eddington sitting out one balmy evening with a girlfriend. She said, "Aren't the stars pretty?" And he said, "Yes, and I'm the only person in the world who knows what makes them shine."

It's a nice story, but it's none too likely. Eddington was a lifelong bachelor, a Quaker, and a workaholic, too busy to have much time for idle philandering. Just as damning for the anecdote, Rudolf Kippenhahn, in his book *100 Billion Suns* (Kippenhahn, 1979), tells exactly the same story—about Fritz Houtermans.

Even Eddington could not say how hydrogen fused to form helium. That insight came ten years later, with the work of Hans Bethe and Carl von Weizsäcker, who in 1938 discovered the "carbon cycle" for nuclear fusion.

However, Eddington didn't have to know how. He had all the information that he needed, because he knew how much energy would be released when four hydrogen nuclei changed to one helium nucleus. That came from the mass of hydrogen, the mass of helium, and Einstein's most famous formula, $E=mc^2$.

Eddington worked out how much hydrogen would have to be converted to provide the Sun's known energy output. The answer is around 600 million tons a second. That sounds like a large amount, but the Sun is a huge object. To keep the Sun shining as brightly as it shines today for five billion years would require that less than eight percent of the Sun's hydrogen be converted to helium.

Why pick five billion years? Because other evidence suggests an age for the Earth of about 4.6 billion years. Nuclear fusion is all we need in the Sun to provide the right time-scale for geology and biology on Earth. More than that, the Sun can go on shining just as brightly for another five billion years, without depleting its source of energy.

But how typical a star is the Sun? It certainly occupies a unique place in our lives. All the evidence, however, suggests that the Sun is a rather normal star. There are stars scores of times as massive, and stars tens of times as small. The Sun sits comfortably in the middle range, designated by astronomers as a G2 type dwarf star, in what is known as the *main sequence* because most of the stars we see can be fitted into that sequence.

The life history of a star depends more than anything else on its mass. That story also started with Eddington, who in 1924 discovered the *mass-luminosity law*. The more massive a star, the more brightly it shines. This law does not merely restate the obvious, that more massive stars are bigger and so radiate more simply because they are of larger area. If that were true, because the mass of a star grows with the cube of its radius, and its surface area like the square of its radius, we might expect to find that brightness goes roughly like mass to the two-thirds power (multiply the mass by eight, and expect the brightness to increase by a factor of four). In fact, the brightness goes up rather faster than the *cube* of the mass (multiply the mass by eight, and the brightness increases by a factor of more than a *thousand*).

The implications of this for the evolution of a star are profound. Dwarf stars can go on steadily burning for a hundred billion years. Massive stars squander their energy at a huge rate, running out of available materials for fusion in just millions of years.

(A word of warning: Don't put into your stories a star that's a thousand times the mass of the Sun, or one-thousandth. The upper limit on size is set by stability, because a contracting ball of gas more than about 90 solar masses will oscillate wildly, until parts of it are blown off into space; what's left will be 90 solar masses or less. At the lower end, below maybe one-twelfth of the Sun's mass, a star-like object cannot generate enough internal pressure to initiate nuclear fusion and should not be called a "star" at all.)

The interesting question is, what happens to massive stars when their central regions no longer have hydrogen to convert to helium? Detailed models, beginning with Fred Hoyle and William Fowler's work on stellar nucleosynthesis in the 1940s, have allowed that question to be answered.

Like a compulsive gambler running out of chips, stars coming to the end of their supply of hydrogen seek other energy sources. At first they find it through other nuclear fusion processes. Helium in the central core "burns" (not chemical burning, but the burning of nuclear fusion) to form carbon, carbon burns to make oxygen and neon and magnesium. These processes call for higher and higher temperatures before they are significant. Carbon burning starts at about 600 million degrees (as usual, we are talking degrees Celsius). Neon burning begins around a billion degrees. Such a temperature is available only in the cores of massive stars, so for a star less than nine solar masses that is the end of the road. Many such stars settle down to old age as cooling lumps of dense matter. Stars above nine solar masses can keep going, burning neon and then oxygen. Finally, above 3 billion degrees, silicon, which is produced in a process involving collisions of oxygen nuclei, begins to burn, and all the elements are produced up to and including iron. By the time that we reach iron, the different elements form spherical shells about the star's center, with the heaviest (iron) in the middle, surrounded by shells of successively lighter elements until we get to a hydrogen shell on the outside.

Now we come to a fact of great significance. *No elements heavier than iron can be produced through this nuclear synthesis process in stars*. Iron, element 26, is the place on the table of elements where nuclear binding energy is maximum. If you try to "burn" iron, fusing it to make heavier elements, you *use* energy, rather than producing it. Notice that this has nothing to do with the mass of the star. It is decided only by nuclear forces.

The massive star that began as mainly hydrogen has reached the end of the road. The final processes have proceeded faster and faster, and they are much less efficient at producing energy than the hydrogen-to-helium reaction. Hydrogen burning takes millions of years for a star of, say, a dozen solar masses. But carbon burning is all finished in a few thousand years, and the final stage of silicon burning lasts only a day or so.

What happens now? Does the star sink into quiet old age, like most small stars? Or does it find some new role?

And one more question. We can explain through stellar nucleosynthesis the creation of every element lighter than iron. But more than 60 elements heavier than iron are found on Earth. If they are not formed by nuclear fusion within stars, where did they come from?

3.2 Stellar endings. We have a star, of ten or more solar masses, running out of energy. The supply provided by the fusion at its center, of silicon into iron, is almost done. In the middle of the star is a sphere of iron "gas" about one and a half times the mass of the sun and at a temperature of a few billion degrees. It acts like a gas because all the iron nuclei and the electrons are buzzing around freely. However, the core density is millions of times that of the densest material found on Earth. Outside the central sphere, like layers of an onion, sit shells of silicon, oxygen and carbon, helium and neon and hydrogen, and smaller quantities of all the other elements lighter than iron.

When the source of fusion energy dries up, iron nuclei capture the free electrons in the iron gas. Protons and electrons combine. The energy that had kept the star inflated is sucked away. The core collapses to become a ball of neutrons.

The near-instantaneous gravitational collapse unleashes a huge amount of energy, enough to blow all the outer layers of the star clear away into space. What is left behind is a "neutron star"—a solid sphere of neutrons, spinning on its axis many times a second, only a few miles across but with a mass as much as the Sun's mass.

When such an object was observed, as a rapidly but regularly varying radio source, it seemed difficult to imagine anything in nature that could explain the signal. The team at Cambridge who discovered the first one in 1967 called it a *pulsar*. They wondered, even if they were reluctant to say so in public, if they had found signals from some alien civilization. When other pulsars were discovered and Thomas Gold proposed that the radio sources were provided by rotating neutron stars, astronomers realized that such a possibility had been pointed out long ago—in 1934, in a prophetic paper by Walter Baade and Fritz Zwicky. The most astonishing thing about the paper was that the neutron itself had been discovered only two years earlier, in 1932.

Could life ever exist on the surface of such a body, with its immense gravitational and magnetic field, and its extreme temperature and dizzying rotation? You might think not, but the novel *Dragon's Egg* (Forward, 1980) explores that wild possibility, as does *Flux* (Baxter, 1993).

And how much is a "huge" amount of energy? When a star collapses and blows up like this, in what is known as a *supernova*, it shines for a time as brightly as a whole galaxy. Its luminosity can temporarily increase by a factor of one hundred billion. If that number doesn't tell you much, try it this way: if a candle in New York were to "go supernova," you would be able to read a newspaper by its light in Washington, D.C.

The explosion of the supernova also creates pressures and temperatures big enough to generate all the elements heavier than iron that could *not* be formed by standard nucleosynthesis in stars. So finally, after a long, complex process of stellar evolution, we have found a place where substances as "ordinary" as tin and lead, or as "precious" as silver, gold, and platinum, can be created.

For completeness, I will point out that there are actually two types of supernova, and that both can produce heavy elements. However, the second kind cannot happen to an isolated star. It occurs only in binaries, pairs of stars, close enough together that material from one of them can be stolen gravitationally by the other.

The star that does the stealing must be a small, dense star of the type known as a *white dwarf*, while its partner is usually a larger, diffuse, and swollen star known as a *red giant*. As more and more matter is stolen from the more massive partner, the white dwarf star shrinks in size, rather than growing. When its mass reaches 1.4 times the mass of the Sun (known as *Chandrasekhar's limit*) it collapses. The result is a huge explosion, with a neutron star left behind as a possible remnant. The outgoing shock wave creates heavy elements, and ejects them from the system along with the rest of the star's outer layers.

If you are thinking of using a supernova as part of a story, note that according to current theory the nearest binary star to us, Alpha Centauri A and B, is not a candidate. I am not discouraging the idea of using such a supernova, since I have just done it myself (*Aftermath*, 1998). The flux of radiation and high-energy particles from an Alpha Centauri supernova can do interesting things to Earth. But you'll need to do some ingenious talking if you want the idea to seem plausible.

Supernovas are rather like nuclear power stations. What they produce is important to us—it is the very stuff of which our own bodies and many of our most valued products are made. We prefer, however, not to have one in our own local neighborhood.

What is the final fate of a star that explodes and becomes a neutron star? That depends on the mass of the part that's left. One possibility is that it remains a neutron star to the end of its life. Another more exotic possibility is that it shrinks further and becomes a black hole. That intriguing option we will describe in the next section, after which we will expand the scale of our exploration.

3.3 Black holes. The story of black holes begins with Albert Einstein and the theory of general relativity.

In 1916, soon after the publication of the field equations in their final form, Karl Schwarzschild produced the first exact solution. Einstein was reportedly quite surprised, because of the complicated nature of the field equations— a set of ten coupled nonlinear partial differential equations. As Einstein wrote to Max Born, twenty years later, "If only it were not so damnably difficult to find rigorous solutions."

The "Schwarzschild solution" gave the gravitational field for an isolated mass, which later became known as the Schwarzschild black hole. At the time, it was considered to be mathematically interesting, but of no physical significance. Soon after Schwarzschild's work, Reissner and Nordstrom solved the general relativity equations for a spherical mass that also carried a charge. It too was regarded with no special interest.

In 1939, Oppenheimer and Snyder studied the collapse of a star under gravitational forces—a situation which certainly *does* have physical significance, since it is a common stellar occurrence.

Two remarks from the summary of their paper are worth quoting: "Unless fission due to rotation, the radiation of mass, or the blowing off of mass by radiation, reduce the star's mass to the order of the sun, this contraction will continue indefinitely." In other words, not only can a star collapse, but if it is heavy enough there is no way that the collapse and contraction can be stopped. And "the radius of the star approaches asymptotically its gravitational radius; light from the surface of the star is progressively reddened, and can escape over a progressively narrower range of angles." This is the first modern picture of a black hole, a body with a gravitational field so strong that light cannot escape from it. We say "modern picture" because John Michell in 1783, and Pierre Laplace in 1798, independently noted that a sufficiently massive body would have an escape velocity from its surface that exceeded the speed of light.

The idea of a "gravitational radius" came straight from the Schwarzschild solution. It is the distance from the center where the reddening of light becomes infinite, and it defines a sphere. Any light coming from inside that sphere can never be seen by an outside observer. Today the surface of the sphere has a variety of names, all defining the same thing: the surface of infinite red shift, the trapping surface, the one-way membrane, and the event horizon. Since the gravitational radius for the Sun is only three kilometers, if it were squeezed down to this size (which will never happen, fortunately, as a result of gravity) conditions inside the collapsed body would be difficult to imagine. The density of matter would be about twenty billion tons per cubic centimeter.

You might suppose that the Oppenheimer and Snyder paper, with its apparently bizarre conclusions, would have produced a sensation. In fact, it aroused little notice. It too was looked on as a mathematical oddity, a result that physicists did not need to take too seriously. The resurgence of interest in the solutions of the equations of general relativity did not take place until after Einstein's death in 1955, and it was one of the leaders of that renaissance, John Wheeler, who in 1958 provided the inspired name for the Schwarzschild solution at the gravitational radius: the *black hole*.

The object described by the Schwarzschild and Reissner/Nordstrom solutions could have a mass, and a charge, and that was all. The next development came in 1963, and it was a big surprise to everyone in the field.

Roy Kerr had been exploring a particular form of the Einstein field equations. The analysis was highly mathematical and seemed to be wholly abstract—until Kerr found that he could produce an exact form of solution. It included the Schwarzschild black hole as a special case, but there was more, another quantity that Kerr was able to associate with *spin*. For the first time, the possibility of a spinning black hole had appeared. It could also, as was shown a couple of years later by Ezra Newman and collaborators, have an associated charge.

From this point on, I am for convenience going to call the charged, spinning Kerr-Newman black hole a *kernel*. It has a number of fascinating properties useful to science fiction writers.

First, since it carries a charge, a kernel can be moved from place to place using electric and magnetic fields. Second, the kernel has associated with it not the single characteristic surface of the Schwarzschild solution (the sphere defined by the gravitational radius), but two. In this case, the surface of infinite red shift is distinct from the event horizon.

To visualize the surfaces, take a hamburger bun and hollow out the inside, enough to let you put a round hamburger entirely within it. For a kernel, the outer surface of the bread (which is a sort of ellipsoid in shape) is the surface of infinite red shift, the "static limit" within which nothing can remain at rest, no matter how hard and efficiently its rocket engine works. Inside the bun, the surface of the meat patty forms a sphere, the "event horizon" from which no light or particle can ever escape to the outside. We can never find out anything about what goes on within the meat patty's surface, so its composition and nature, like that of many hamburgers, must remain a complete mystery. For a kernel, the bun and patty surfaces touch only at the north and south poles of the axis of rotation (the top and bottom centers of the bun). A really interesting region, however, lies between these two surfaces. It is called the *ergosphere*, and it has a most unusual property, pointed out in 1969 by Roger Penrose (yes, the same Penrose as in Chapter 2—he is a highly versatile and creative individual, who has made major contributions to relativity theory and other fields).

Penrose showed that it is possible for a particle to dive in toward the kernel from outside, split in two when it is inside the ergosphere, and then have part of it ejected to the exterior in such a way that the piece has more *total* energy than the particle that went in. If we do this, we have *extracted energy* from the black hole.

Note that this must be a kernel, a *spinning* black hole, not a Schwarzschild black hole. The energy that we have gained comes from the rotational energy of the hole itself.

If the kernel starts out with only a little spin energy, we can use the energy-extraction process in reverse, to provide more rotational energy. We will refer to that as "spinning up" the kernel. "Spin down" is the opposite process, the one that extracts energy.

One other property of a kernel will prove useful later. Every kernel (but not a Schwarzschild black hole) possesses a "ring singularity." It appears possible to remain far enough from the singularity to avoid destruction by tidal forces, but close enough to take advantage of peculiar aspects of space-time there. This is discussed further in Chapter 9.

Since it can be proved that a black hole has as properties only mass, charge, spin, and magnetic moment, and the last one is fixed completely by the other three, that seems to say all that can be said about kernels. This result, that all black holes are completely defined by three constants, is a theorem that is often stated in the curious form, "A black hole has no hair."

That was the situation until 1974, when Stephen Hawking produced a result that shocked everyone. In perhaps the biggest surprise in all black hole history, he proved that *black holes are not black*.

This calls for some explanation. General relativity and quantum theory were both developed in this century, but they have never been combined in a satisfactory way. Physicists have known this and been uneasy about it for a long time. In attempting to move toward what John Wheeler referred to as "the fiery marriage of general relativity with quantum theory," Hawking studied quantum mechanical effects in the vicinity of a black hole. He found that particles and radiation can (and must) be emitted from the hole.

The smaller (and therefore less massive) the hole, the faster the rate of radiation. Hawking was able to relate the mass of the black hole to a temperature, and as one would expect, a "hotter" black hole pours out radiation and particles faster than a "cold" one. For a black hole the size of the Sun, the associated temperature is far lower than the background temperature of the Universe (the 2.7 Kelvin background radiation). Such a black hole receives more energy than it emits, so it will steadily increase in mass. However, there is no rule of nature that says a black hole has to be big and massive. For a black hole of a few billion tons (the mass of a small asteroid) the temperature is so high, at ten billion degrees, that the black hole will radiate itself away to nothing in a gigantic and rapid burst of radiation and particles. Furthermore, a spinning black hole will preferentially radiate particles that decrease its spin, while a charged black hole will prefer to radiate charged particles that reduce its overall charge.

These results are so strange that in 1972 and 1973 Hawking spent a lot of time trying to find the mistake in his own calculations. Only when he had performed every check that he could think of was he finally forced to accept the conclusion: black holes are not black after all; and the smallest black holes are the least black.

We have discussed the properties of kernels, without asking the crucial question: Do they exist?

For a while, it was thought that very small black holes, weighing only a hundredth of a milligram, might have been created in the Big Bang. The Hawking radiative process showed that any of those, if they ever existed, would have gone long since. Big black holes, however, seem not only possible, but inevitable. If the core of a collapsing star is massive enough (more than about three times the mass of our Sun), then after the star explodes to a supernova, Oppenheimer and Snyder's results apply. The remnant star is forced to collapse without limit, and no force in the universe is powerful enough to stop it.

Black holes, if they exist at all, ought therefore to be

common throughout the universe, perhaps enough to make a sizable contribution to the missing mass needed to close it (see Chapter 4). However, some people object to the very idea of black hole existence. Associated with them is a singularity—an infinity—that no one has been able to explain away, and singularities are generally regarded as evidence that a theory has something wrong with it. Einstein himself was reported to consider black holes as a "blemish to be removed from his theory by a better mathematical formulation."

Until that better mathematical formulation comes along, black holes are an acceptable part of theoretical physics; but what is the experimental evidence for them?

We have a problem. A black hole, unless it is small (the mass of, say, a small asteroid) will not radiate measurable energy. Also, we know of no way that a black hole less than about three solar masses can form. Black holes are therefore, by definition, not directly visible. Their existence, like the existence of quarks, depends not on observing them, but on the role they play in simplifying and explaining other observations.

A black hole's presence must be detected by indirect effects. For example, matter falling into a black hole will be ripped apart and give off a powerful radiation signal; but so will matter that falls into a neutron star. Distinguishing between the two is a subtle and difficult problem. One of the early and best candidates for a solar-sized black hole is the source known as Cygnus X-1.

Very large black holes probably lie at the heart of many galaxies, and are the mechanism that powers quasars. It is also possible to regard the whole universe as a black hole, within which we happen to live; but these are conjectures, not established facts. In view of Einstein's comment, maybe any other possible explanation is to be preferred.

Despite lack of final proof of their existence, black holes form a valuable weapon in the writer's arsenal. In fact, they are so accepted a feature of the science fiction field that they can be introduced without further explanation, like an alien or a faster-than-light drive. Black holes, of various sizes and properties, can be found in hundreds of stories.

I will mention just a handful, so you will not be tempted to write a classic story that already exists: "The Hole Man," by Larry Niven (Niven, 1973); *Imperial Earth*, by Arthur C. Clarke (Clarke, 1975); *Beyond the Blue Event Horizon*, by Frederik Pohl (Pohl, 1980); and *Earth*, by David Brin (Brin, 1990). All of these employ the Schwarzschild black hole. I like better the spinning, rotating black hole. With my preferred name for them, a kernel (for **Ker**r-**New**man black hole), they are used in all the stories in *One Man's Universe*, and in *Proteus Unbound* (Sheffield, 1983, 1989).

CHAPTER 4
Physics in the Very Large

4.1 Galaxies. The ancient astronomers, observing without benefit of telescopes, knew and named many of the stars. They also noted the presence of a hazy glow that extends across a large fraction of the sky, and they called it the *Milky Way*. Finally, those with the most acute vision had noted that the constellation of Andromeda contained within it a much smaller patch of haze.

The progress from observation of the stars to the explanation of hazy patches in the sky came in stages. Galileo started the ball rolling in 1610, when he examined the Milky Way with his telescope and found that he could see huge numbers of stars, far more than were visible with the unaided eye. He asserted that the Milky Way was nothing more than stars, in vast numbers. William Herschel carried this a stage farther, counting how many stars he could see in different parts of the Milky Way, and beginning to build towards the modern picture of a great disk of billions of separate stars, with the Sun well away (30,000 light-years) from the center.

At the same time, the number of hazy patches in the sky visible with a telescope went up and up as telescope power increased. Lots of them looked like the patch in Andromeda. A dedicated comet hunter, Charles Messier, annoyed at constant confusion of hazy patches (uninteresting) with comets (highly desirable) plotted out their locations so as not to be bothered by them. This resulted in the *Messier Catalog*: the first and inadvertent catalog of galaxies.

But what were those fuzzy glows identified by Messier? The suspicion that the Andromeda and other galaxies might be composed of stars, as the Milky Way is made up of stars, was there from Galileo's time. Individual stars cannot usually be seen, but only because of distance. The number of galaxies, though, probably exceeds anything that Galileo would have found credible. Today's estimate is that there are about a hundred billion galaxies in the visible universe—roughly the same as the number of individual stars in a typical galaxy. Galaxies, fainter and fainter as their distance increases, are seen as far as our telescopes can probe.

In most respects, the distant ones look little different from the nearest ones. But there is one crucial difference. And it tells us something fundamental about the whole universe.

4.2 The age of the universe. Galaxies increase in numbers as they decrease in apparent brightness, and it is natural to assume that these two go together: if we double the distance of a galaxy, it appears one-quarter as bright, but we expect to see four times as many like it if space is uniformly filled with galaxies.

What we would not expect to find, until it was suggested by Carl Wirtz in 1924 and confirmed by Edwin Hubble in 1929, is that more distant galaxies appear *redder* than nearer ones.

To be more specific, particular wavelengths of emitted light have been shifted towards longer wavelengths in the fainter (and therefore presumably more distant) galaxies. The question is, what could cause such a shift?

The most plausible mechanism, to a physicist, is called the *Doppler Effect*. According to the Doppler Effect, light from a receding object will be shifted to longer (redder) wavelengths; light from an approaching object will be shifted to shorter (bluer) wavelengths. Exactly the same thing works for sound, which is why a speeding police car's siren seems to drop in pitch as it passes by.

If we accept the Doppler effect as the cause of the reddened appearance of the galaxies, we are led (as was Hubble) to an immediate conclusion: the whole universe must be *expanding*, at a close to constant rate, because the red shift of the galaxies corresponds to their faintness, and therefore to their distance.

Note that this does not mean that the universe is expanding into some other space. There is no other space. It is the whole universe—everything there is—that has grown over time to its present dimension.

From the recession of the galaxies we can draw another conclusion. If the expansion proceeded in the past as it does today, there must have been a time when everything in the whole universe was drawn together to a single point. It is logical to call the period since then, *the age of the universe*. The Hubble galactic red shift allows us to calculate that length of time. The universe seems to be between ten and twenty billion years old.

We have here a truly remarkable result: observation of the faint agglomerations of stars known as galaxies has led us, very directly and cleanly, to the conclusion that we live in a universe of finite and determinable age. A century ago, no one would have believed such a thing possible.

The recession of the galaxies also, in a specific sense, says that we live in a universe of finite and determinable size. For since, according to relativity theory, nothing can move faster than light, the "edge of the universe" is the distance at which the recession velocity of the galaxy is light speed. Nothing can come to us from farther away than that. There could be anything out there, anything at all, and we would never know it.

Answering one question inevitably leads to another: Can we say anything more about the other "edge" of the universe, the time that defines its beginning?

One approach is to use our telescopes to peer farther into space. When we do this, we are also looking farther back in *time*. If a galaxy five billion light-years away sent light in our direction, that radiation has been on the way for five billion years. Therefore, if we can look far enough out, at galaxies eight or even ten billion light-years, we will be observing the early history of the universe.

There is one big built-in assumption here: the observed red shift has to be associated with a velocity of recession, and therefore with distance. One mysterious class of objects with large red shifts has led some people to question that assumption. These are the *quasars* (a contraction of quasi-stellar radio source, or quasi-stellar object).

Quasars are characterized by their large red shifts, which suggests they are very distant, and by their brightness, which means they have a very high intrinsic luminosity at least comparable with a galaxy. And they are *small*. We know this not because they fail to show a distinct disc, which is not surprising at their presumed distances, but because their variations in light patterns take place over such short periods that we know they cannot be more than a few light-hours across. That is no more than the size of our own solar system.

The big question is, how can something so small be so bright?

The only mechanism that anyone has been able to suggest is of a massive black hole (a hundred million times the mass of our sun, or more) into which other matter is falling. This proves an extraordinarily efficient way of creating lots of energy. Almost half the mass of the in-falling matter can be converted to pure radiation. If one or two stars a year were to fall into a monster black hole, that would be enough to power the quasar.

There are, however, reputable scientists who do not believe this explanation at all. According to them, quasars are not at galactic distances. They are much closer, much smaller and less bright, and the red shift of their light is due to some other cause.

What other cause? We will mention one possibility in Chapter 13. Meanwhile, we assume the validity of the Big Bang model.

4.3 Early days. "Oh, call back yesterday," said Salisbury, in Shakespeare's *Richard II*. "Bid time return."

What was the universe like, ten or twenty billion years ago, when it was compressed into a very small volume?

Surprisingly, we can deduce a good deal about those early days. The picture is a coherent one, consistent with today's ideas of the laws of physics. It also, quite specifically, says something about the formation of elements during those earliest times.

Like much of twentieth-century physics, the story begins with Albert Einstein. After he had developed the general theory of relativity and gravitation, he and others used it in the second decade of this century to study theoretical models of the universe. Einstein could construct a simple enough universe, with matter spread through the whole of space. What he could not do was make it sit still. The equations insisted that the model universe either had to expand, or it had to contract.

To make his model universe stand still, Einstein introduced in 1917 a new, and logically unnecessary, "cosmological constant" into the general theory. With that, he could build a stable, static universe. He later described the introduction of the cosmological constant, and his refusal to accept the reality of an expanding or contracting universe, as the biggest blunder of his life. More on this in Section 4.11.

When Hubble's work showed the universe to be expanding, Einstein at once recognized its implications. However, he himself did not undertake to move in the other direction, and ask about the time when the contracted universe was far more compact than it is today. That was done by a Belgian, Georges Lemaître. Early in the 1930s Lemaître went backwards in time, to the period when the whole universe was a "primeval atom." In this first and single atom, everything was squashed into a sphere only a few times as big as the Sun, with no space between atoms, or even between nuclei. Later scientific thought suggests that the primeval atom was far smaller yet. As Lemaître saw it, this unit must then have exploded, fragmenting into the atoms and stars and galaxies and everything else in the universe that we know today. He might justifiably have called it the *Big Bang*, but he didn't. That name was coined by Fred Hoyle (the same man who did the fundamental work on nucleosynthesis) in 1950. It is entirely appropriate that Hoyle, whose career has been marked by colorful and imaginative thinking, should have named the central event of modern cosmology. And it is ironic that Hoyle himself, as we will see in Chapter 13, denies the reality of the Big Bang.

Lemaître did not ask about the composition of the primeval atom. It might be thought that the easiest assumption is that everything in the universe was already there, much as it is now. But that cannot be true, because as we go back in time, the universe had to be hotter as well as more dense. Before a certain point, atoms as we know them could not exist; they would be torn apart by the intense radiation that permeated the whole universe.

The person who did worry about the composition of the primeval atom was George Gamow. In the 1940s, he conjectured that the original stuff of the universe was nothing more than densely packed neutrons. Certainly, it seemed reasonable to suppose that the universe at its outset had no net charge, since it seems to have no net charge today. Also, a neutron left to itself will decay radioactively, to form an electron and a proton. One electron and one proton form an atom of hydrogen; and even today, the universe is predominantly hydrogen atoms. So neutrons could account for most, if not all, of today's universe.

If the early universe was very hot and very dense and all hydrogen, some of it ought to have fused and become helium, carbon, and other elements. The question, *How much of each?*, was one that Gamow and his student, Ralph Alpher, set out to answer. They calculated that about a quarter of the matter in the primeval universe should have turned to helium, a figure consistent with the present composition of the oldest stars.

What Gamow and Alpher could not do, and what no one else could do after them, was make the elements heavier than helium. In fact, Gamow and colleagues proved that heavier element synthesis did *not* take place. It could not happen very early, because in the earliest moments, elements would be torn apart by energetic radiation. At later times, the universe expanded and cooled too quickly to provide the needed temperatures.

Heavier element formation has to be done in *stars*, during the process known as stellar nucleosynthesis. The failure of the Big Bang to produce elements heavier than helium confirms something that we already know, namely, that the Sun is much younger than the universe. Sol, at maybe five billion years old, is a second, third, or even fourth generation star. Some of the materials that make up Sun and Earth derive from older stars that ran far enough through their evolution to produce the heavier elements by nuclear fusion and in supernovas.

4.4 All the way back. We are now going to run time backward toward the Big Bang. (Note: this section draws heavily from the book *The First Three Minutes* [Weinberg, 1977]. I strongly recommend the original.)

Where should we start the clock? Well, when the universe was smaller in size, it was also hotter. In a hot enough environment, atoms as we know them cannot hold together because high-energy radiation rips them apart as fast as they form. A good time to begin our backward running of the clock is the time when atoms could form and persist as stable units. Although stars and galaxies would not yet exist, at least the universe would be made up of familiar components: hydrogen and helium atoms.

Atoms formed, and held together, somewhere between half a million and a million years after the Big Bang. Before that time, matter and radiation interacted continuously. Afterward, radiation could not tear matter apart as fast as it was formed. The two "de-coupled," or nearly so, became quasi-independent, and went their separate ways. Matter and radiation still interacted (and do so to this day), but more weakly. The temperature of the universe when this happened was about 3,000 degrees. Ever since then, the expansion of the universe has lengthened the wavelength of the background radiation, and thus lowered its temperature. The cosmic background radiation discovered by Penzias and Wilson, at 2.7 degrees above absolute zero, is nothing more than the radiation at the time when it decoupled from matter, now grown old.

Continuing backwards: before atoms could form, helium and hydrogen nuclei and free electrons could exist; but they could not combine to make atoms, because radiation broke them apart. The form of the universe was, in effect, controlled by radiation energetic enough to prevent atom formation. This situation held from about three minutes to one million years A.C. (After Creation).

If we go back before three minutes A.C., radiation was even more dominant. It prevented the build-up even of helium nuclei. As noted earlier, the fusion of hydrogen to helium requires hot temperatures, such as we find in the center of stars. But fusion cannot take place if it is *too* hot. For helium nuclei to form, three minutes after the Big Bang, the universe had to "cool" to about a billion degrees. All that existed before this time were electrons (and their positively charged forms, positrons), neutrons, protons, neutrinos, and radiation.

Until three minutes A.C., you might think that radiation controlled events. Not so. As we proceed backwards and the temperature of the primordial fireball continues to increase, we reach a point where the temperature is so high (above ten billion degrees) that large numbers of electron-positron pairs are created from pure radiation. That happened from one second to 14 seconds A.C. After that, the number of electron-positron pairs decreased rapidly, because less were being generated than were annihilating themselves and returning to pure radiation. When the universe "cooled" to ten billion degrees, neutrinos also decoupled from other forms of matter.

We have a long way to go, physically speaking, to the moment of creation. As we continue backwards, temperatures rise and rise. At a tenth of a second A.C., the temperature of the universe is 30 billion degrees. The universe is a soup of electrons, protons, neutrons, neutrinos, and radiation. However, as the kinetic energy of particle motion becomes greater and greater, effects caused by differences of particle mass are less important. At 30 billion degrees, an electron easily carries enough kinetic energy to convert a proton into the slightly heavier neutron. In this period free neutrons are constantly decaying to form protons and electrons, but energetic proton-electron collisions undo their work by remaking neutrons.

The clock keeps running backward. The important time intervals become shorter and shorter. At one ten-thousandth of a second A.C., the temperature is one thousand billion degrees. The universe is so small that the density of matter, everywhere, is as great as that in the nucleus of an atom (about 100 million tons per cubic centimeter; a fair-sized asteroid, at this density, fits in a thimble). The universe is a sea of quarks, electrons, neutrinos, and energetic radiation.

We can go farther, at least in theory, to the time, 10^{-35} seconds A.C., when the universe went through a super-rapid "inflationary phase," growing from the size of a proton to the size of a basketball in about 5×10^{-32} seconds. We can even go back to a time 10^{-43} seconds A.C. (termed the *Planck time*), when according to a class of theories known as *supersymmetry* theories, the force of gravity decoupled from everything else, and remains decoupled to this day.

The times mentioned so far are summarized in TABLE 4.1 (p. 98). Note that all these times are measured from the moment of the Big Bang, so $t=0$ is the instant that the universe came into being.

TABLE 4.1 displays one inconvenient feature. Everything seems to be crowded together near the beginning, and major events become farther and farther apart in time as we come closer to the present. This is even more apparent when we note that the origin of the solar system, while important to us, has no cosmic significance.

Let us seek a change of time scale that will make important events more evenly spaced on the time line. We make a change of the time coordinate, defining a new time, T, by $T=\log(t/t_N)$, where t_N is chosen as 15 billion years, the assumed current age of the universe.

That produces TABLE 4.2 (p. 98). All the entries in it are negative, since we have been dealing so far only with past times. However, the entries for important events, in cosmological terms, are much more evenly spaced in T-time.

We will return to TABLE 4.2 later. Note, however, that we cannot get all the way to the Big Bang in T-time, since that would correspond to a T value of minus infinity. However, a failure to reach infinite pressure and temperature is no bad thing. In T-time, the Big Bang happened infinitely long ago.

The time transformation that we made to T-time has no physical motivation. It gives us a convenient time scale for spacing past events, in terms of a familiar function, but there is no reason to think it will be equally convenient in describing the future.

A value of $T = +60.7$, which is as far ahead of the present on the T-time scale as the Planck time is behind us, corresponds to a time of 7.5×10^{70} years from now.

Does the future of the universe admit such a time? We shall see.

At this point, however, I want to pause and ask, does it make any sense to go back so far? If we try to press "all the way back" to zero time, we find ourselves faced with a singularity, a time when matter density and temperature tend to infinity. The appearance of infinity in a physical theory is one good way of knowing that there is something wrong—not with the universe, but with the theory. The most likely problem is that physical laws derived under one set of conditions cannot be applied to grossly different conditions. However, it is also possible that the theory itself is too naive.

In either case, we are already far away from the scientific mainland, well into science fiction waters. We are certainly beyond the realm of the physical laws that we can test today. We are at this stage no more plausible than Archbishop Ussher, convinced that he had pinned down the time of creation.

More to the point, does the early history of the universe make any difference to *anything* today?

Oddly enough, it does. The early history was crucial in deciding the whole structure of today's universe. Let us see why.

4.5 The missing matter. The universe is expanding. Almost every cosmologist today agrees on that. Will it go on expanding forever, or will it one day slow to a halt, reverse direction, and fall back in on itself in a "Big Crunch"? Or is the universe perhaps poised on the infinitely narrow dividing line between expansion and ultimate contraction, so that it will increase more and more slowly in size and finally (but after infinite time) stop its growth?

We also ought to mention still another possibility, that the universe *oscillates*, going through endless phases of expansion followed by contraction. This idea, known as *kinematic relativity*, was developed by E.A. Milne (not, please, to be confused with A.A. Milne), but it has now fallen from favor.

The thing that chooses among the three main possibilities is the total amount of mass in the universe; or rather, since we do not care what form the mass takes, and mass and energy are totally equivalent, the future of the universe is decided by the total mass-energy content per unit volume.

If the mass-energy is too big, the universe will end in the Big Crunch. If it is too small, the universe will fly apart forever. Only in the Goldilocks situation, where the mass-energy is "just right," will the universe ultimately reach a "flat" condition.

The amount of matter needed to stop the expansion of the universe is not large, by terrestrial standards. It calls for only three hydrogen atoms per cubic meter.

Is there that much available?

If we estimate the mass and energy from visible material in stars and galaxies, we find a value nowhere near the "critical density" needed to make the universe finally flat. If we arbitrarily say that the critical mass-energy density has to be unity to end the expansion after infinite time, we observe a value of only 0.01.

There is evidence, though, from the rotation of galaxies, of more "dark matter" than visible matter. It is not clear what this dark matter is—black holes, very dim stars, clouds of neutrinos—but when we are examining the future of the universe, we don't care. All we worry about is the amount. And that amount, from galactic dynamics, could be at least ten times as much as the visible matter. Enough to bring the density to 0.1, or possibly even 0.2. But no more than that.

One might say, all right, that's it. There is not enough matter in the universe to stop the expansion, by a factor of about ten, so we have confirmed that we live in a forever-expanding universe.

Unfortunately, that is not the answer that most cosmologists would really like to hear. The problem comes because the most acceptable cosmological models tell us that if the density is as much as 0.1 today, then in the past it must have been much closer to unity. For example, at one second A.C., the density would have had to be within one part in a million billion of unity, in order for it to be 0.1 today. It would be an amazing coincidence if, by accident, the actual density were so close to the critical density.

Most cosmologists therefore say that, today's observations notwithstanding, the density of the universe is *exactly* equal to the critical value. In this case, the universe will expand forever, but more and more slowly.

The problem, of course, is then to account for the matter that we don't observe. Where could the "missing matter" be that makes up the other nine-tenths of the universe?

There are several candidates. And now, I should point out, we are very much into science fiction territory.

One suggestion is that the universe is filled with energetic ("hot") neutrinos, each with a small but nonzero mass (as mentioned earlier, the neutrino is usually assumed to be massless). Those neutrinos would be left over from the very early days of the universe, so we are forced back to studying the period soon after the Big Bang. However, there are other problems with the Hot Neutrino theory, because if they are the source of the mass that stops the expansion of the universe, the galaxies, according to today's models, should not have developed as early as they did.

What about other candidates? Well, the class of theories already alluded to and known as supersymmetry theories require that as-yet undiscovered particles ought to exist.

There are *axions*, which are particles that help to preserve certain symmetries (charge, parity, and time-reversal) in elementary particle physics; and there are *photinos*, *gravitinos*, and others, based on theoretical super-symmetries between particles and radiation. These candidates are slow-moving (and so considered "cold") but some of them have substantial masses. They too would have been around soon after the Big Bang. These slow-moving particles clump more easily together, so the formation of galaxies could take place earlier than with the hot neutrinos. We seem to have a better candidate for the missing matter—except that no one has yet observed the necessary particles. Neutrinos are at least known to exist!

Supersymmetry, in the particular form known as *superstring theory*, offers one other possible source of hidden mass. This one is easily the most speculative. Back at the time, 10^{-43} seconds A.C., when gravity decoupled from everything else, a second class of matter may have been created that interacts with normal matter and radiation only through the gravitational force. We can never observe such matter in the usual sense, because our observational methods, from ordinary telescopes to radio telescopes to gamma ray detectors, all rely on *electromagnetic* interaction with matter. The "shadow matter" produced at the time of gravitational decoupling lacks any such interaction with the matter of the familiar universe. We can determine its existence only by the gravitational effects it produces; which, of course, is exactly what we need to "close the universe." Unfortunately, the invocation of shadow matter takes us back to such an early time that if we are sure of anything, it is that the universe then was unrecognizably different from the way that it is today.

I used shadow matter in a story ("The Hidden Matter of McAndrew," Sheffield, 1992). However, I took care to be suitably vague about its properties.

4.6 The end of the universe. "When I dipped into the Future far as human eye could see," said Tennyson in the poem "Locksley Hall." Writing in 1842 he did pretty well, foreseeing air warfare and universal world government. We can go a long way beyond that.

Let's start with the "near-term" future. We can model mathematically the evolution of our own sun. In the near-term (meaning in this case the next few billion years) the results are unspectacular. The sun is a remarkably stable object. It will simply go on shining, becoming slowly brighter. Five billion years from now it will be twice its present diameter, and twice as bright. Eventually, however, it will begin to deplete its stock of hydrogen. At that point it will not shrink as one might expect, but begin to balloon larger and larger. Eight billion years in the future, the sun will be two thousand times as bright, and it will have grown so big (diameter, a hundred million miles) that its sphere will fill half our sky. The oceans of Earth will long since have evaporated, and the land surface will be hot enough to melt lead.

That far future Sun, vast, stationary and dim-glowing in the sky of an ancient Earth, was described by H.G. Wells in one of the most memorable scenes in science fiction (*The Time Machine*, 1895). The details are wrong—his future Earth is cold, not hot—but the overall effect is incredibly powerful. If you have not read it recently, it well repays rereading.

In studying the long-term future of the sun, we have as an incidental dealt with the future of the Earth. It will be incinerated by the bloated sun, which by that time will be a red giant. The sun, as its energy resources steadily diminish even further, will eventually blow off its outer layers of gas and shrink to end its life, ten billion years from now, as a dense white dwarf star not much bigger than today's Earth.

None of this should be a problem for humanity. Either we will be extinct, or long before five billion years have passed we will have moved beyond the solar system. We can, if we choose, go to sit around a smaller star. It will be less prodigal with its nuclear fuel, and we can enjoy its warmth for maybe a hundred billion years. By that time the needs of our descendants will be quite unknowable.

However, before that time something qualitatively different may have happened to the universe. Just possibly, it will have ended. We know that the universe is open, closed, or flat, but no one knows which. We must examine all three alternatives.

4.7 The Big Crunch. We begin with the case of the closed universe, which is in many ways the least appealing. It has to it a dreadful feeling of finality—though it is not clear why a human being, with a lifetime of a century or so, should be upset by events maybe fifty to a hundred billion years in the future.

The Big Crunch could happen as "soon" as 50 billion years from now, depending on how much the average mass-energy of the universe exceeds the critical amount. We know from observation that the mass-energy density is not more than twice the critical density. In that limiting case we will see about 17.5 billion more years of expansion, followed by 32.5 billion years of collapse. A smaller mass-energy density implies a longer future.

Not surprisingly, T-time is inappropriate to describe this future. The logarithm function has a singularity at $t=0$, but nowhere else. An appropriate time for the closed universe contains not one singularity ($T=-\infty$, the Big Bang), but two ($T=-\infty$, the Big Bang, and $T=+\infty$, the Big Crunch). As the universe approaches its end, the events that followed the Big Bang must appear in inverse order. There will come a time when atoms must disappear, when helium splits back to hydrogen, when electron/positron pairs appear, and so on.

A reasonable time transformation for the closed universe is $T_c=\log(t/(C-t))$, where C is the time, measured from the Big Bang, of the Big Crunch.

TABLE 4.3 (p. 99) shows how this transformation handles significant times of the past and future. In this case, we have chosen $T_c=0$ as the midpoint in the evolution of the universe, equally far from its beginning and its end. For past times, the values are very similar to those obtained with T-time. For future times close to the Big Crunch, T-time and T_c-time are radically different.

As the universe is collapsing to its final singularity T_c-time is rushing on to infinity, but the hands of the T-time clock would hardly be moving.

T_c is a plausible time to describe the evolution of a closed universe. When t tends to zero, T_c tends to minus infinity, and when t tends to C, T_c tends to plus infinity. Thus both end points of the universe are inaccessible in T_c-time. The transformation is symmetric about the "midpoint" of the universe, t=C/2. This does not mean, as is sometimes said, that time will "run backwards" as the Universe collapses. Time continues to run forward in either t-time or T_c-time, from the beginning of the Universe to its end. Note also that T_c has no real values, and hence no meaning, for times before the Big Bang or after the Big Crunch.

Since the collapse applies to the whole universe, there is no escape—unless one can find a way to leave this universe completely, or modify its structure. I dealt with both those possibilities in the novel *Tomorrow and Tomorrow* (Sheffield, 1997).

4.8 At the eschaton. I want to mention another aspect of the end of the universe, something that appears only in the case where it is closed. Consider the following statement:

> *The existence of God depends on the amount of matter in the universe.*

That is proposed, as a serious physical theory, by Frank Tipler. It was the subject of a paper (Tipler, 1989) and a later book (Tipler, 1994). Both concern the "eschaton." That is the final state of all things, and it therefore includes the final state of the universe.

Tipler argues that certain types of possible universes allow a physicist to deduce (his own term is *prove*) the ultimate existence of a being with omnipresence, omniscience, and omnipotence. This being will have access to all the information that has ever existed, and will have the power to resurrect and re-create any person or thing that has ever lived. Such a being can reasonably be called God.

The universe that permits this must satisfy certain conditions:

1) The universe must be such that life can continue for infinite subjective time.

2) Space-time, continued into the future, must have as a boundary a particular type of termination, known as a c-boundary.

3) The necessary c-boundary must consist of a single point of space-time.

Then, and only then, according to Tipler, God with omnipresence, omniscience, and omnipotence can be shown to exist.

Conditions 2) and 3) are satisfied only if the universe is *closed*. It cannot be expanding forever, or even asymptotically flat, otherwise the theory does not work. The choice, open or closed, depends as we already noted on the mass-energy density of the universe.

The definition of "omnipotent" now becomes extremely interesting. Would omnipotence include the power to avoid the final singularity, by changing the universe itself to an open form?

I like to think so, and in *Tomorrow and Tomorrow* I took that liberty.

When the question of missing matter and the closed or open universe was introduced, it seemed interesting but quite unrelated to the subject of religion. Tipler argues that the existence of God, including the concepts of resurrection, eternal grace, and eternal life, depends crucially on the *current* mass-energy density of the universe.

We already noted the surprising way in which the observation of those remote patches of haze, the galaxies, showed that the universe began a finite time ago. That was a striking conclusion: Simple observations today defined the far past of the universe.

Now we have a still stranger notion to contemplate. The search for exotic particles such as "hot" neutrinos and "cold" photinos and axions will tell us about the far future of the universe; and those same measurements will have application not only to physics, but to theology.

4.9 Expansion forever. Suppose that the universe is open rather than closed. Then it will expand forever.

Freeman Dyson was the first to analyze this situation (Dyson, 1979). First, all ordinary stellar activity, even of the latest-formed and smallest suns, will end. That will be somewhat less than a million billion (say, 10^{14}) years in the future. After that it is quiet for a while, because everything will be tied up in stellar leftovers, neutron stars and black holes and cold dwarf stars.

Then the protons in the universe begin to decay and vanish.

That requires a word of explanation. A generation ago, the proton was thought to be an eternally stable particle, quite unlike its cousin, the unstable free neutron. Then a class of theories came along that said that protons too may be unstable, but with a vastly long lifetime. If these theories are correct, the proton has a finite lifetime of at least 10^{32} years. In this case, as the protons decay all the stars will finally become black holes.

The effect of proton decay is slow. It takes somewhere between 10^{30} and 10^{36} years before the stellar remnants are all black holes. Note that on this time scale, everything that has happened in the universe so far is totally negligible, a tick at the very beginning. The ratio of the present age of the universe to 10^{36} years is like a few nanoseconds compared with the present age of the universe.

In terms of T-time, the stellar remnants collapse to form black holes between $T = 19.8$ and $T = 25.8$. The T-transformation still does pretty well in describing the open universe.

Long after the protons are all gone, the black holes go, too. Black holes evaporate, as we saw in Chapter 3. Today, the universe is far too hot for a black hole of stellar mass to be able to lose mass by radiation and particle production. In another 10^{64} years or so that will not be true. The ambient temperature of the expanding universe will have dropped and dropped, and the black holes will evaporate. Those smaller than the Sun in mass will go first, ones larger than the Sun will go later; but eventually all, stars, planets, moons, clouds of dust, everything, will turn to radiation.

In this scenario, the universe, some 10^{80} years from now, will be an expanding ocean of radiation, with scattered within it a possible sprinkling of widely-separated electron-positron pairs.

The idea of proton decay is controversial, so we must consider the alternative. Suppose that the proton is not an unstable particle. Then we have a rather different (and far longer) future for the universe of material objects.

All the stars will continue, very slowly, to change their composition to the element with the most nuclear binding energy: iron. They will be doing this after some 10^{1600} years.

Finally (though it is not the end, because there is no end) after somewhere between 10 to the 10^{26} and 10 to the 10^{76} years, a time so long that I can find no analogy to offer a feel for it, our solid iron neutron stars will become black holes. Now our T-time scale also fails us. A t-time of 10 to the 10^{26} years corresponds to $T=10^{26}$, itself a number huge beyond visualization.

Is this the end of the road? No. The black holes themselves will disappear, quickly on these time scales. The whole universe, as in the previous scenario, becomes little more than pure radiation. This all-encompassing bath, feeble and far-diluted, is much too weak to permit the formation of new particles. A few electron-positron pairs, far apart in space, persist, but otherwise radiation is all.

TABLES 4.4, 4.5 and 4.6 (pp. 99–100) show the calendar for the future in "normal" t-time, for the closed and open universes with the unstable and stable proton. Time is measured from today, rather than the beginning of the universe.

4.10 Life in the far future. There is something a little unsatisfactory about the discussion so far. A universe, closed or open, without anyone to observe it, feels dull and pointless.

What are the prospects for observers and conscious participants, human or otherwise? We will certainly not equate "intelligence" with "humanity," since over the time scales that we have encountered, the idea that anything like us will exist is remotely improbable.

Let us note that, on the cosmological scale, life as we know it on Earth has a respectable ancestry. Life emerged quite early in this planet's lifetime, about three and a half billion years ago, so life is now about a fourth as old as the universe itself.

Land life appeared much later, 430 million years ago for simple plants. The first land animals came along a few tens of millions of years later. Mammals have existed for maybe 225 million years, and flowering plants about a hundred million. Recognizable humans, with human intelligence, appeared a mere three or four million years ago. We are upstarts, in a universe where ordinary turtles have been around, essentially unchanged in form and function, for a couple of hundred million years. Perhaps that is why we lack the calm certainty of the tortoise.

Humans have a short past, but we could have a long future. We have already taken care of the "near-term" future. The Earth should remain habitable (unless we ourselves do something awful to it) for a few billion years. After that we can head for a dwarf star, and be comfortable there for another thirty to a hundred billion years. Dwarf stars shine dimly, but a planet or free-space colony in orbit a few million miles away from one will find more than enough energy to support a thriving civilization; and of all the stars in the universe, the inconspicuous, long-lived dwarf stars appear to form the vast majority.

Earth, of course, will be gone—unless perhaps our descendants, displaying a technology as far beyond ours as we are beyond the Stone Age, decide to take the home planet along on their travels for sentimental reasons.

If we are to consider longer time scales, beyond thirty billion years, we must distinguish between the cases of an open and a closed universe.

In the universe of the Big Crunch it seems obvious that life and intelligence cannot go on forever. The future contains a definite time at which everything in existence will be compressed to a single point of infinite pressure and temperature. If we continue to measure time in the usual way, life can exist for a finite time only. However, we have already noted that in T_c-time, even the Big Crunch is infinitely far away. Although the transformation that we introduced seemed like a mere mathematical artifice, it can be shown that there is enough time (and available energy) between now and the Big Crunch to think an infinite number of thoughts. From that point of view, if we work with *subjective* time, in which life survives long enough to enjoy infinite numbers of thoughts, that will be like living "forever" according to one reasonable definition. It is all a question of redefining our time coordinates.

The open universe case has no problem with available time, but it does have a problem with available energy. In the far future our energy sources will become increasingly diluted and distant.

Dyson has also analyzed this situation (personal communication, Dyson, 1992). He has examined the possibility of continued life and intelligence for the case of an asymptotically flat spacetime, where the universe sits exactly on the boundary of the open and closed cases. I have not seen the details of his analysis, and to my knowledge they have not been published. Here, however, are his conclusions.

First, hibernation will be increasingly necessary. The fraction of time during which a thinking entity can remain "conscious" must become less, like $t^{-1/3}$. Also, the thinking rate must decrease, so that "subjective time" will proceed more slowly, like $t^{-1/3}$. To give an example of what this implies, one million billion years from now you will be able to remain awake for only ten years out of each million. And during those ten years, you will only be able to do as much thinking as you can do now in one hour. There will be no more "lightning flashes of wit." Instead it will all be Andrew Marvell's "vaster than empires and more slow." All thought must be "cool calculation."

The good news is that you have an indefinitely long time available, so that you can eventually think an infinitely large number of thoughts.

Curiously enough, in an ultimately flat universe an infinite number of thoughts can be thought with the use of only a finite amount of energy. That's just as well, because in such a universe free energy becomes less and less easy to come by as time goes on.

4.11 Complications from the cosmological constant. Recent observations (1999) suggest that the universe is not only open, but is expanding *faster* over time. If these observations hold up, they will eliminate the Big Crunch possibility. They also force me, in a late addition to the text, to say a little more about something that I had hoped to avoid: the "cosmological constant."

Since the presence of matter can only slow the expansion of the universe, what could possibly speed up expansion?

Let's go back again to the early days of the theory of general relativity, when Einstein noted that his equations permitted the introduction of a single extra variable, which he called the *cosmological constant*, generally denoted by Λ. It was not that the equations *required* the added term, they merely permitted it; and certainly the equations appeared more elegant without Λ. However, by including the cosmological constant, Einstein was able to produce a universe that neither expanded nor contracted.

Then in 1928, Hubble offered evidence that distant galaxies were receding and the whole universe was expanding. At that point, almost every physicist would have been more than happy to throw out the cosmological constant, as unnecessary. Unfortunately, like a fairy-tale evil spirit, Λ proved much easier to raise than to banish. No one could prove that Λ was necessarily equal to zero. It was known that Λ must be small, since a non-zero Λ produces a "pressure" in space-time, encouraging the expansion of the universe regardless of the presence of matter. Thus the rate of expansion of the universe sets an upper limit on the possible value of Λ.

Even so, the cosmological constant was felt to be somehow "unphysical." Kurt Gödel, the famous logician, added weight to that idea. He became interested in general relativity when he was Einstein's colleague at the Princeton Institute for Advanced Study, and in 1949 he produced a strange solution of the Einstein field equations with a non-zero Λ. In Gödel's solution the whole universe has an inherent rotation, which the real universe according to all our measurements does not. This was regarded as evidence that a non-zero cosmological constant could lead to weird results and was therefore unlikely.

However, weirdness for the universe is certainly permitted. We have seen enough evidence of that on the smallest scale, in the quantum world. And now we have the possibility that remote galaxies, flying apart from each other faster and faster under the pressure of a non-zero Λ, offer evidence on the largest scale that once again, in the words of J.B.S. Haldane, "the universe is not only queerer than we suppose, it is queerer than we *can* suppose."

TABLE 4.1

Event	time, t
Gravity decouples	10^{-43} seconds
Inflation of universe	10^{-35} seconds
Nuclear matter density	0.0001 seconds
Neutrinos decouple	1 second
Electron/positron pairs vanish	30 seconds
Helium nuclei form	3.75 minutes
Atoms form	1 million years
Galaxy formation begins	1 billion years
Birth of solar system	10 billion years
Today	15 billion years

TABLE 4.2
$T = \log(t/t_N)$, where t_N is chosen as 15 billion years.

Event	T
Big Bang	$-\infty$
Gravity decouples	-60.7
Inflation of universe	-52.7
Nuclear matter density	-21.7
Neutrinos decouple	-17.7
Electron/positron pairs vanish	-16.2
Helium nuclei form	-15.3
Atoms form	-4.2
Galaxy formation begins	-1.2
Birth of solar system	-0.2
Today	0

TABLE 4.3
$$T_c = \log(t/\log(C-t))$$

Event	T_c
Big Bang	$-\infty$
Gravity decouples	-61.31
Inflation of universe	-52.31
Nuclear matter density	-22.31
Neutrinos decouple	-18.31
Electron/positron pairs vanish	-16.83
Helium nuclei form	-15.95
Atoms form	-4.81
Galaxy formation begins	-1.81
Birth of solar system	-0.74
Today	-0.52
"Halfway" point (32.5 billion yrs)	0
Helium dissociates	$+15.95$
Electron/positron pairs form	$+16.83$
Neutrinos couple	$+18.31$
Gravity couples	$+61.31$
End point (65 billion years)	$+\infty$

TABLE 4.4
Closed Universe.

Event	t (billions of years)
Today	0
Sun becomes red giant	5
Halfway point, expansion ceases	17
Most dwarf stars cease to shine	30
Big Crunch	50

TABLE 4.5
Open Universe, Unstable Proton.

Event	t (years)
Today	0
Sun becomes red giant	5 billion
Most dwarf stars cease to shine	30 billion
All stellar activity ceases	10^{14}
Stellar remnants become black holes	10^{30}–10^{36}
Black holes evaporate	10^{64}
Radiation only	10^{80}

TABLE 4.6
Open Universe, Stable Proton.

Event	t (years)
Today	0
Sun becomes red giant	5 billion
Most dwarf stars cease to shine	30 billion
All stellar activity ceases	10^{14}
Stars are iron neutron stars	10^{1600}
Neutron stars form black holes	> 10 to the 10^{26}
Radiation only	10 to the 10^{76}

CHAPTER 5
The Constraints of Chemistry

There is another way to distinguish physics from chemistry. Physics is needed in describing the world of the very small (atoms and down) and the very large (stars and up). Chemistry works with everything in between, from molecules to planets. So although physics is vital to us (where would we be without gravity and sunlight?), chemical processes largely control our everyday lives.

An exception to this rough rule has been created in the last century, as a result of human activities. Lasers, nuclear power, and all electronics from computers to television derive from the subatomic world of physics.

5.1 Isaac Asimov and the Timonium engine. Isaac Asimov was a famous science fiction writer, justly proud of the breadth of his knowledge. He wrote books on everything that you can think of, inside and outside science.

In the nineteenth century, the following jingle was made up about Benjamin Jowett, a famously learned Oxford Don:

"I am Master of this college,
 What I don't know, isn't knowledge."

About Asimov, we might offer this variation:

"I am science fiction's guru,
 What I don't know, don't say you do."

This doesn't actually rhyme, but it makes a point. Asimov knew *lots*. So when he, on a panel at a science fiction convention in Baltimore, heard one of the other speakers refer to a spaceship whose "engines were powered by timonium," and he noticed that all the audience laughed in a knowing way, he was quite put out.

Only later did he learn that this was a purely local reference. *Timonium* sounds like the name of an element, similar to titanium or plutonium, but it is actually a suburb of Baltimore. There was little chance that Asimov would know about it, and the other speaker relied on that fact.

Why, though, was Asimov so sure that timonium *wasn't* an element, perhaps a newly-discovered one between, say, titanium and chromium? Simply because there is already an element, vanadium, between titanium and chromium; and *there cannot be any others*. Titanium has atomic number 22, vanadium 23, chromium 24. If you say in a story that someone has found another element in there, is it much like saying that you have discovered a new whole number between eight and nine.

In fact, although the normal ordering of the elements corresponds to their atomic weights, their numbering is just the number of electrons that surround the nucleus. Hydrogen has one, helium has two, lithium has three, beryllium has four, and so on. Nothing has, or can have, a fractional number of electrons.

Moreover, the electrons are not buzzing around the nucleus at random, they are structured into "electron shells," each of which holds a specific number of electrons. Chemical reactions involve electrons only in the outermost shells of an atom, and this decides all their chemical properties. The shells have been given names. Proceeding outward from the nucleus, we have K, L, M, N, O, P, and Q. Easy to remember, but of no physical significance.

The innermost shell can hold only two electrons. Hydrogen has one electron, so it has a place for one more, or alternatively we might think that it has one electron to spare. Hydrogen can take an electron from another atom, or it can share its single electron with something else. Helium, on the other hand, has two electrons. That makes a filled shell, and in consequence helium has no tendency to share electrons. Helium does not react appreciably with anything. If you write a story in which a race of aliens are helium-breathers, you'd better have a mighty unusual explanation.

Other elements with closed shells are neon (complete shell of two plus complete shell of eight), argon, (shell of two plus shell of eight plus shell of eight), krypton, xenon, and radon. You may recognize these as the "noble gases" or "inert gases." All resist combination with other elements. Radon, element number 86, decays radioactively, but fairly slowly, over a period of days. Radioactive decay, since it involves the *nucleus* of the atom, is by our definition a subject for physics rather than chemistry.

Atoms with one space left in a filled shell also combine very readily, particularly when that "hole" can be matched up with an extra electron in some other element which has one too many for a filled shell. Elements with one electron less than a filled shell include hydrogen, fluorine, chlorine, bromine, iodine, and astatine. These elements are all strongly reactive, and collectively they are known as "halogens." Halogen means "salt-producing," for the good reason that these elements, in combination, all produce various forms of salts. The last element in the halogen list, astatine, element number 85, is also unstable. In a few hours it decays radioactively. However, the properties of astatine—while it lasts—are very similar chemically to the properties of iodine.

Atoms with one electron too many for a filled shell include the elements lithium, sodium, potassium, rubidium, and cesium. Known collectively as the *alkali metals*, they combine readily with the halogens (and many other elements) and form stable compounds. Note that we can choose to regard hydrogen as a halogen, an alkali metal, or both.

The actual number of electrons in each shell is decided by quantum theory. However, long before quantum theory had been dreamed up and before anyone knew of electrons, the elements had been formed into groups in terms of their general chemical properties. This was done by Mendeleyev, who about 1870 had developed a "periodic table" of the elements. It is still in use today, suitably extended by elements discovered since Mendeleyev's time—discovered, in large part, because he had used the periodic table to predict that they ought to be there.

We repeat, for emphasis: the chemical properties of an element are *completely* decided by the number of electrons in its outer shell only. There is no scope for adding new elements "in the cracks" of the periodic table, and little scope for new chemical properties beyond those known today.

Can we find a way around that hindrance, and give the writer some room to maneuver?

We can. The so-called "natural" elements begin with hydrogen, atomic number 1, and end with uranium, atomic number 92. Uranium is itself radioactive, so over a long enough period (billions of years) it decays to form lighter elements. Heavier elements than uranium are not impossible to make, but they are unstable. In a short time—for elements with high enough atomic numbers, small fractions of a second—they decay and become some other, lighter element.

If we could just make stable elements *heavier* than uranium, or anything else known to our laboratories today, a whole new field of chemistry would open up. These new "transuranic elements" could have who-knows-what interesting properties.

Now, as heavier and heavier elements are created beyond uranium, their radioactive decay to other elements normally takes less and less time. It seems hopeless to look for new stable elements. However, there is one ray of hope. The neutrons and protons that make up the atomic nucleus form, like the electrons outside it, "shells." When the number of neutrons in a nucleus has certain values (known as "magic numbers"), the corresponding element is unusually stable. Similarly, extra stability is achieved when the number of protons in a nucleus has "magic" values. If a nucleus has the right number of protons (usually written Z) *and* the right number of neutrons (usually written N), it is known as "doubly magic," and is correspondingly doubly stable.

Magic numbers, computed from the shell theory of the nucleus (and generally agreeing with experiment), are 2, 8, 20, 28, 50, 82, 114, 126, and 184. The theoretical calculations provide the higher values, but these are not seen in naturally occurring elements which end at uranium with $Z=92$. In principle, doubly-magic numbers would occur with any pair of these magic numbers, such as $Z=20$ and $N=8$. In practice, in every heavy nucleus, the number of neutrons is greater than or equal to the number of protons. Also, a nucleus is unstable if N exceeds Z (known as the "*neutron excess*") by a large factor. Given these two rules, we might expect nuclei of extra stability for $Z=2$, $N=2$; $Z=8$, $N=8$; $Z=20$, $N=20$; $Z=28$, $N=28$; and so on.

What we find in practice is that $Z=2$, $N=2$ is helium, and the nucleus is highly stable. $Z=2$, $N=8$ is not stable at all, because the excess of neutrons over protons is too large. $Z=8$, $N=8$ is oxygen, and it is very stable. So is $Z=20$, $N=20$ (calcium, stable), and $Z=82$, $N=126$ (lead, also stable).

We now see a possibility that doubly-magic, extra-stable elements might exist with $Z=114$ or $Z=126$, and a suitably high neutron number, N, of 184. Experiments so far have not led to any such elements, but the existence of an "island of stability" somewhere between element 114 and element 126 is a suitable offshore location for science fictional use.

Even more interesting is the possibility that humans might someday discover a way to *stabilize* naturally radioactive materials against decay. We know of no way to do this at the moment, but we can argue that it might be possible, with an analogy provided by Nature. A free neutron, left to itself, will usually decay in a quarter of an hour to yield a proton and an electron. Bound within a nucleus, however, the same neutron acts as a stable particle. The helium nucleus, two protons and two neutrons, is one of the most stable structures known. Perhaps, by embedding a super-heavy nucleus from the island of stability within some larger structure, we can prevent its decay for an indefinitely long period. The super-heavy transuranic elements might then share a property of the quark, that while we understand their properties, we never actually observe them.

5.2 The limits of strength. The strength and flexibility of a material, anything from chalk to cheese, depend on the bonds between its atoms and molecules. Since interactions at the atomic level take place only between the electrons in the outer shell of atoms, the strength of those bonds is decided by them.

The density and mass of a material, on the other hand, is decided mainly by the atomic nucleus. For every electron in an atom there must be a proton in the nucleus, plus possible neutrons, and the neutron and proton each outweigh the electron by a factor of almost two thousand.

We thus have an odd contrast:

Strength: determined by outer electrons.

Weight: determined by nucleus.

Using these two facts, without any other information whatsoever, we can reach a conclusion. The strongest materials, for a given weight, are likely to be those which have the most outer electrons, relative to the total number of electrons, and the least number of neutrons (which are purely wasted weight) relative to the number of protons.

The elements with the most electrons in the outer shell relative to the total number of electrons are hydrogen and helium. As we know already, helium is a poor candidate for any form of strong bond. Also, whereas the hydrogen nucleus has no neutrons, the helium nucleus has as many neutrons as protons. We conclude, on theoretical grounds, that the strongest possible material for its weight ought to be some form of hydrogen. Of course, hydrogen is a gas at everyday temperatures, but that should not deter us.

TABLE 5.1 (p. 122) shows the strength/weight ratio of different materials. It confirms our theoretical conclusion. Hydrogen, in solid form, ought to be the strongest "natural" material—once we have produced it.

We have added to the table "below the line" a couple

of extra items with a decidedly science fictional flavor. Muonium is like hydrogen, but we have replaced the single electron in the hydrogen atom with a muon, 207 times as massive. The resulting atom will be 207 times smaller than hydrogen, and should have correspondingly higher bonding strength.

Muonium, considered as a building material, is not without its problems. The muon has a lifetime of only a millionth of a second. In addition, because the muon spends a good part of its time close to the proton of the muonium atom, there is a fair probability of spontaneous proton-proton fusion.

Positronium takes the logical final step of getting rid of the wasted mass of the nucleus completely. It replaces the proton of the hydrogen atom by a positron, an electron with a positive charge. Positronium, like muonium, has been made in the lab, but it too is unstable. It comes in two varieties, depending on spin alignments. Parapositronium decays in a tenth of a nanosecond. Orthopositronium lasts a thousand times as long, a full ten-millionth of a second.

As in the case of our transuranic elements, we rely upon future technology to find some way to stabilize them.

5.3 The production of energy. Burning and the strength of materials do not sound to have much in common, but they are alike in this: they both depend on the relationship of the outer electrons of atoms to each other. We use the word "burning" to mean not only the combination of another element with oxygen, but to describe the combination of any elements by chemical means. We also include the very rapid burning that would normally be called an explosion.

We would like to know the maximum possible energy that can be obtained by chemical combination of a fixed total amount of any materials. This information will prove particularly important for space travel.

One way to find out the energy content of typical fuels using standard oxygen burning is to look up their "caloric value." We can also determine, for any particular material, how much weight of oxygen is needed per unit of fuel. Hence, looking ahead to the needs of Chapter 8, we can calculate how fast the total burned material would travel if the heat produced were entirely converted to energy of motion.

The result does not give a wide range of answers for many of the best known fuels. Pure carbon (coal) gives an associated velocity of 4.3 kilometers a second. Ethyl alcohol leads to the same value. Methane and ethane have almost identical values, at 4.7 kms/sec. The highest value is achieved with hydrogen, at 5.6 kms/sec.

These values are more than can be achieved in practice, because all the energy produced does not go into kinetic energy. Otherwise, the combustion products would be at room temperature (assuming they started there—liquid hydrogen and liquid oxygen, which together make an excellent rocket fuel, must be stored at several hundred degrees below zero). Also, from thermodynamic arguments, conversion of heat energy to motion energy can never be one hundred percent efficient.

Why assume that oxygen must be one component in the fuel? Only because oxygen is in plentiful supply in our atmosphere. However, it turns out that oxygen is a good choice for another reason: it combines fiercely with other elements. It is also relatively light (atomic weight, 16). When we examine other combinations of elements, the only one better than a hydrogen-oxygen combination is hydrogen and fluorine. Fluorine, the next element in the atomic table to oxygen, is a halogen, which as we have already noted means it is strongly reactive. When we burn hydrogen and fluorine, and convert all the energy to motion, we find a velocity of 5.64 kms/sec.

This is a very small gain over hydrogen/oxygen, and there are other disadvantages to using fluorine. The result of the combustion of hydrogen and oxygen is water, as user-friendly a compound as one can find. The combination of hydrogen and fluorine, however, yields hydrofluoric acid, a most unpleasant substance. Among other things, it dissolves flesh. Release it into the atmosphere in a rocket launch, and you will have the Environmental Protection Agency jumping all over you.

The disadvantages of the hydrogen-fluorine combination exceed its advantages. Hydrogen and oxygen provide the best fuel in practice.

Are there other options to improve performance? One possibility is suggested by something that we noted in discussing the strength of materials. Chemical reactions, like chemical bonds, are decided by the interaction of the electrons that surround the nucleus of an atom. However, the weight of an atom is provided almost completely by the nucleus. We must have protons, to make the atom electrically neutral. But if we want to accelerate a material to high speed, the neutrons in the nucleus are just dead weight.

We would achieve a higher final speed from combustion if we replaced normal oxygen by some lighter form of it. Such an idea is not impossible, because many elements have something known as *isotopes*. An isotope of an element has the usual number of protons, but a different number of neutrons. For example, hydrogen comes in three isotopic forms: H^1 is "normal" hydrogen, the familiar element with one proton and one electron; H^2 has one proton, one neutron, and one electron. It is a stable form, with its own name, *deuterium*. Finally, H^3 has one proton, two neutrons, and one electron. It is slightly unstable, decaying radioactively over a period of years.

However, this takes us in the wrong direction. We are interested in isotopes with fewer neutrons than usual, not more. Oxygen has a total of eight isotopes. The most common form of the atom, O^{16}, has eight protons, eight neutrons, and eight electrons. The four heavier isotopes, O^{17}, O^{18}, O^{19}, and O^{20}, all have more neutrons and are of no interest. There is, however, an isotope O^{13}, with only five neutrons. If we use this in place of normal oxygen, the maximum speed associated with hydrogen-oxygen combustion increases from 5.6 kms/sec to 6.15 kms/sec. Unfortunately, O^{13} decays radioactively in a fraction of a second; however, O^{14} is longer-lived, and its use gives a maximum speed of 5.95 kms/sec. The similar use of a lighter isotope of fluorine, F^{17}, gives a speed of 5.8 kms/sec.

It seems fair to say that 6 kms/sec provides an absolute

upper limit for an exhaust speed generated using chemical fuels. Putting on our science fiction hats, can we see any possible way to do better than this?

Chemical combustion involves two atoms, originally independent, that combine to share one or more of their electrons. Also, as we have seen, neutrons take no part in this process. They just provide useless weight. We would therefore expect the ideal chemical fuel would be one in which no neutrons are involved, and in which the energy contribution from the electrons is as large as possible.

The best conceivable situation should thus involve only hydrogen (H^1, a single proton with no neutron), and obtain the largest possible energy release involving an electron. This occurs when a free electron approaches a single proton, to form a neutral hydrogen atom. The energy release for this case is well-known. It is termed the *ionization potential* for hydrogen, and it is measured in a particular form of unit known as an *electron volt*. One electron volt (shortened to eV) is the energy required to move an electron a distance of one centimeter in an electric field of one volt. That sounds like a very strange choice of unit, but it proves highly convenient in the atomic and nuclear world, where most of the numbers we have to deal with are nicely expressed in electron volts. The masses of nuclear particles, recognizing the equivalence of mass and energy, are normally written in eV or MeV (million electron volts) rather than in kilograms or some other inconveniently large unit (an electron masses only 9.109310^{-31} kilograms).

The ionization potential of hydrogen is 13.6 eV. The mass of an electron is 0.511 MeV, and of a proton 938.26 MeV. Knowing these facts and nothing else, we have enough to calculate the maximum speed obtained when neutral hydrogen forms from a proton and a free electron. Write the kinetic energy of the product as $\frac{1}{2}mv^2$, where m is the mass of electron plus proton, and so equals 938.77 MeV. To convert this from the form of an energy to a mass, we invoke $E=mc^2$ from Chapter 2, and divide by c^2. The energy provided by the electron is 13.6 eV. Equating these two, we have $\frac{1}{2} \times 938.7731,000,000 \ (v/c)^2 = 13.6$. Using $c = 300,000$ kms/sec and solving for v, $v = 51.06$ kms/sec.

This is the absolute, ultimate maximum velocity we can ever hope to achieve using chemical means. It is also surely unattainable. To do better, or even as well in practice, we must turn to the realm of physics and the violent processes of the subatomic world.

Orders of magnitude more energy are available there. To give an example: the ionization potential of hydrogen is 13.6 eV, so this is the energy released when a free electron and a free proton combine to form a hydrogen atom. The nuclear equivalent, combining two protons and two neutrons to form a helium nucleus, yields 28 MeV—over two million times as much.

5.4 Organic and inorganic: building an alien. Those grandparents of modern chemistry, the alchemists of five hundred years ago, had a number of things on their wish list. One, however, dominated all the others. The alchemists sought the *philosopher's stone*, able to convert base metals to gold.

Claiming to be able to transmute metals, and failing, had stiff penalties in the fifteenth and sixteenth centuries. Marco Bragadino was hanged by the Elector of Bavaria, William de Krohnemann by the Margrave of Bayreuth, David Benther killed himself before he could be executed by Elector Augustus of Saxony, and Marie Ziglerin, one of the few female alchemists, was burned at the stake by Duke Julius of Brunswick. Frederick of Wurzburg had a special gallows, gold painted, for the execution of those who promised to make gold and failed.

The inscription on a gibbet where an alchemist was hanged read: "Once I knew how to fix mercury, and now I am fixed myself."

Today, we know that the philosopher's stone is a problem not of chemistry, but of physics. The transmutation of elements was first shown to be possible in the early 1900s, by Lord Rutherford, when he demonstrated how one element could change to another by radioactive decay, or through bombardment with subatomic particles. For this achievement, the physicist Rutherford—ironically, and to his disgust—was awarded the 1908 Nobel Prize. In chemistry.

Next on the alchemists' list was the *universal solvent*, capable of dissolving any material. That problem has vanished with the advance of chemical understanding and knowledge of the structure of compounds. Today, we have solvents for any *given* material. *Aqua regia*, known to the alchemists, is a mixture of one part concentrated nitric acid with three parts concentrated hydrochloric acid. It will dissolve most things, including gold. Hydrofluoric acid is only a moderately strong acid (unlike, say, hydrochloric acid) but it will dissolve glass. However, even today we have no single, universal solvent.

The old question when considering the problem still exists: If you did make the solvent, what would you keep it in?

The third item on the alchemist's list of desirable discoveries was the *elixir of life*. This would, when drunk or perhaps bathed in, confer perpetual youth. The quest for it not only occupied the alchemists in their smoky laboratories, but sent explorers wandering the globe. Juan Ponce de Leon was told by Indians in Puerto Rico that he would find the Fountain of Youth in America. He sailed west and discovered not the fountain but Florida, a region today noted less for perpetual youth than for perpetual old age. Cosmetic surgery, aerobics, and vitamins notwithstanding, the elixir eludes us still. But as we will see in the next chapter, we may be on the threshold of a breakthrough.

The three alchemical searches are often grouped together, but the third one is fundamentally different from the other two. The first pair belong totally to the chemical world. The elixir of life crosses the borderline, to the place where chemistry interacts with an organism— humans, in this case—to produce a desired effect.

Five hundred years ago, people were certainly doing this in other ways. That is what medical drugs are all about. However, there was a strong conviction that living organisms were not just an assembly of chemicals. Plants and animals were thought to be basically *different* from inorganic forms. They contained a "vital force" unique to living things.

It was easy to hold this view when almost every substance found in the human body could not be made in the alchemist's retorts. The doubts began to grow when chemists such as the Frenchman Chevreul were unable to detect any differences between certain fats occurring in both plants and animals. The key step was taken in 1828, when Friedrich Wöhler was able to synthesize urea, a substance never before found outside a living organism (actually, this is not quite true; urea had been prepared in 1811 by John Davy, but not recognized).

From that beginning, the chemists of the nineteenth and twentieth centuries one by one produced, from raw materials having nothing to do with plants or animals, many of the sugars, proteins, fats, and vitamins found in the bodies of animals and humans. With their success, it slowly became clear why vitalism had seemed reasonable for so long. The molecules of simple compounds are made up of a few atoms; carbon dioxide, for example, is one atom of carbon and two of oxygen. Copper sulfate is one atom of copper, one of sulfur, and four of oxygen. By contrast, many of the molecules of our bodies contain thousands of atoms. The difference between the molecules of living things and those of nonliving materials is largely one of scale.

More than that, biological compounds depend for their properties very much on the way they are constructed. Two big molecules can have exactly the same number of atoms of each element, but because of their different connecting structure they have totally different properties (such molecules, like in composition yet unalike in structure, are called *isomers*). Wöhler's success with urea was due at least in part to the fact that it is, as biological molecules go, simple and small, containing only eight atoms. In fact, urea is not so much a building block of a living organism, as a convenient way of dealing with the excretion of ammonia, an undesirable by-product of other reactions.

Chemists noticed one other thing. The big molecules of biochemistry all seem to contain *carbon*. In fact, the presence of carbon is so strong an indicator of organic matter, the terms "organic" and "inorganic" in chemistry have nothing to do with the origin of a material. Organic chemistry is, quite simply, the chemistry of materials that contain carbon. Inorganic chemistry is everything else. The distinction is not quite foolproof. Few people would refer to the study of a very simple molecule, such as carbon dioxide or methane (CH_4), as organic chemistry. They reserve the term for the study of substantial molecules that contain carbon. "Biogenic" is a better term than "organic" to describe the chemistry of living things, but today the latter is used to refer to both biogenic and carbon chemistry.

Why is carbon so important? What is there about carbon that makes it so different, so able to aid in the construction of giant molecules? This question is particularly important if we want to devise alien chemistries. Is it absolutely necessary that alien life-forms, no matter their star or planet of origin, be based on carbon?

Let us return to the shell model of the atom. Each shell around the nucleus can hold a specific number of electrons, and chemical reactions involve only those electrons in the outermost shell.

The innermost shell can hold two electrons. The next will hold another eight, for a total of ten. Atoms with spaces in a filled shell match up such electron "holes" with the extra electrons of other substances outside a filled shell.

Now note the curious situation of carbon. It has six electrons surrounding its nucleus. Thus, it has four extra electrons beyond the two of the first filled shell. On the other hand, it is four electrons short of filling a second shell; thus it has both four extra electrons, and four "holes" to be filled by other electrons. This "ambivalence" (a chemical joke; literally, two valences or strengths) of carbon makes it capable of elaborate and complex combinations with other elements. It is also, as we will see later in this chapter, capable of making elaborate and surprising combinations with itself.

Is carbon unique? Again we consider the shell model.

The third shell can hold another eight electrons. Thus, an element with four electrons more than needed to fill the second shell, namely, fourteen, will be four electrons short of filling the third shell. Like carbon, it will have four extra electrons, and at the same time four electron holes to be filled.

Element fourteen is *silicon*. We have been led to it, by a very natural and simple logic, as a substance with the same capacity as carbon to form complex molecules. It can serve as the basis for a "silico-organic" chemistry, the stuff of aliens.

There will of course be differences between carbon-based and silicon-based life forms. For example, carbon dioxide is a gas at room temperature. Silicon dioxide is a solid with several different forms (quartz, glass, and flint are the most familiar), and remains solid to high temperatures. These differences are an interesting challenge to the writer. Just don't use the carbon/silicon analogy blindly. An alien who breathes in oxygen and excretes silicon dioxide is not impossible, but does deserve some explanation.

5.5 Building a horse. There is no real difference between the chemistry of life and the chemistry of the inanimate world, other than complexity. "Vitalism" is dead. This simple fact demolishes the idea of a "food pill," found in rather old and rather bad science fiction.

The food pill is an aspirin-sized object that taken twice a day, with water, supplies all the body's needs. Apart from the sheer unpleasantness of the idea (no more pizza, no more veal *cordon bleu*, no more ice cream), it won't work. The body runs just like any other engine, burning organic fuel to produce energy and waste products. We have seen that there are definite limits to the energy produced by chemical reactions. A couple of small pills a day is not enough, no matter how efficiently they are used. To get by on a food pill, the human body would first have to go nuclear.

Chemistry and biochemistry are subject to identical physical laws. If we like, we can regard biochemistry as no more than a branch of all chemistry. Conversely, we can use the chemistry of living organisms to perform the functions of general chemistry.

To take one example, the marine organism known as a tunicate has a curious ability to concentrate vanadium from sea water. If we want vanadium, it makes sense to use this "biological concentrator" (which also provides its own fuel supply and makes its own copies). In science fiction, it is quite permissible to presume that an organism can be developed to concentrate any material at all— gold, silver, uranium, whatever the story demands. It is also reasonable to assume that the principle employed by the tunicate will eventually be understood, so that we can make a "vanadium concentrator" along the same lines, but without the tunicate.

The analogy between chemical and biological systems is not always a fruitful one. When I was pondering the question of the most efficient chemical rocket fuel, I noted that energy was always wasted in heating the exhaust. A hot exhaust jet does not deliver more thrust than the same mass expelled cold. Greater efficiency would therefore be obtained if the exhaust could somehow be at room temperature.

That sounds impossible, but we and all other animals have in our bodies large numbers of specialized proteins known as *enzymes*. The purpose of an enzyme is to control chemical reactions, making them proceed at much lower temperatures than usual, or faster or slower.

Suppose we build an "enzymatic engine" in which the chemical fuels are combined to release as much energy as usual, but the temperature remains low? We might then have a better rocket for launches to space.

I soon realized that the idea would not work, because as we point out in Chapter 8, the whole idea in using rockets for a launch is to burn the fuel *as fast as possible*. There is no way to achieve a fast burn, yet avoid a temperature rise in the fuel's combustion products. Forget that, then. But what about an enzymatic engine for other purposes? Say, to power a vehicle that moves on the ground. For such a use, slow and steady fuel consumption is preferable to a single, giant, near-explosion. There would be other advantages, too. The intense heat generated when we burn fuels such as gasoline is a big factor in a vehicle's wear and tear. The heat also generates nitrogen oxides, which are a serious form of air pollution.

Let us imagine, then, a method of ground transportation, powered by some kind of slow-burning enzymatic engine that can operate at close to room temperature. Such a device would have numerous uses, and be free of environmental problems.

I was expounding on this idea with some fervor when a more hardheaded friend of mine pointed out that I seemed to be designing a horse.

5.6 Fullerenes: a chemical surprise. Textbooks on inorganic chemistry for the past couple of centuries have stated, without a hint of doubt, that carbon occurs in two and only two elementary forms: diamond, and graphite. In diamond, the carbon atoms form tetrahedra, triangular pyramids with one carbon atom at each vertex and one in the center. This is a strong and stable configuration, so diamond is famously hard. In graphite, the carbon atoms form hexagons with an atom at each vertex, and the hexagons line up as layers of flat sheets. Since the sheets are not strongly coupled with each other, graphite is famously slippery and a well-known lubricant.

The discovery in 1985 of a third elementary form of carbon was a shock in two different ways. First, the existence of the third form could have been predicted, or at least conjectured, since the middle of the eighteenth century. In fact, its existence was suggested in 1966, as a piece of near-whimsical speculation, by a columnist in the *New Scientist* magazine. No one took any notice. Second, and almost a disgrace to a self-respecting chemist, the third form is not at all hard to make. In fact, it had been around, waiting to be discovered, in every layer of soot produced by a hot carbon fire. Every time you light a candle, at least some of the soot will be this new and previously unknown form of carbon.

I said, *this* new form of carbon, but actually there is a family of them. The simplest form, C_{60}, is sixty carbon atoms arranged in a round hollow shape involving 12 pentagons and 20 hexagons. Technically, this form is called a truncated icosahedron, but the name is neither suggestive nor catchy. However, the structure looks exactly like a tiny soccer ball.

Leonhard Euler, the great Swiss mathematician, studied the possible geometry of closed spheroidal structures more than two hundred years ago, and proved that while they must have exactly 12 pentagons, the number of hexagons may vary. And vary they do. Continuing the sporting motif, the next simplest form, C_{70}, is an oblong spheroid of 12 pentagons and 25 hexagons that closely resembles a rugby ball. And after that there are carbon molecules with 76, 84, 90, and 94 atoms, and still bigger versions that form hollow closed tubes. All of these are known by the generic name of "fullerenes," or if they are round, "buckyballs." The form with 60 atoms, C_{60}, is the simplest, most stable, and most abundant form, with C_{70} in second place. Not surprisingly, C_{60} was the first form to be discovered.

So how was it discovered? Not, as one might think, by direct observation. The C_{60} molecule is less than a millionth of a millimeter across (about 7×10^{-10} meters), but it is big enough to be seen using a scanning tunneling microscope. It wasn't, though. It was found by a very curious and apparently improbable route. A British chemist, Harold Kroto, was studying how carbon-rich stars might lead to the production of long chains of carbon molecules in open space. In the United States, at the Houston campus of Rice University, American chemists Robert Curl and Richard Smalley had suitable lab equipment to simulate the carbon-rich star environment and see what might be happening.

The team did indeed find evidence of a variety of carbon clusters, but as the carbon vapor was allowed to condense, everything else seemed to fade away except for a 60-atom cluster, and, much less abundant, a 70-atom cluster. It seemed that there must be a very stable form of carbon with just 60 atoms, and another, rather less stable, with 70 atoms.

At this point, the team faced a problem. Carbon is

highly reactive. If the cluster had the form of a flat sheet, like graphite, it ought to have free edges which would latch on to other carbon atoms, and so grow rapidly in size. The only way around that would be if the structure could somehow close in on itself, and tie up all the loose ends.

The research team was guided at that point not by the eighteenth-century mathematical researches of Euler, but by the geodesic dome idea of Buckminster Fuller. That, too, is a closed structure of pentagons and hexagons. With faith that a closed 60-atom sphere like a geodesic dome was the only plausible structure for the cluster, the researchers went ahead and named it "buckminsterfullerene." They did have the grace to apologize for such a mouthful of a name, and it was quickly shortened to "fullerenes" when it was realized that there was not one but a multitude of molecules.

The first fullerenes were produced in minute quantities. Research on them was therefore difficult. Then in 1990 a German team discovered a shockingly simple production method. By burning a graphite rod electrically, the resulting soot contained a substantial percentage of C_{60}. Combining this with the suitable use of a benzene solvent, an almost-pure mixture of fullerenes was formed. Now anyone who wants fullerenes for research can easily buy them. And they are doing so, in ever-increasing numbers. The buckyball was named "Molecule of the Year" by *Science* magazine in 1991, and today the most frequently cited chemistry papers *all* seem to be on the subject of fullerenes. The 1996 Nobel Prize in chemistry went to Robert Curl, Richard Smalley, and Harold (now Sir Harold) Kroto.

One natural question is, all right, so fullerenes exist, and they are of scientific interest. But what are they *good for*, apart from winning Nobel Prizes? Potentially, many things. Because they are hollow, buckyballs can be used to trap other atoms inside them and to provide miniature "chemical test sites." They are phenomenally robust and stable, and could be the basis for materials stronger than anything we have today. They have been proposed as nanotechnology building blocks. They are already being used to improve the growth of diamond films. And they have interesting properties and potential as super-conductors.

The best answer to the question, though, is that it is too soon to say. Like lasers in 1965, five years after the first one was built, fullerenes seem to be a solution waiting for a problem. And like lasers, fullerenes will almost certainly become enormously valuable techno-logical tools in the next thirty years.

5.7 A burning home: the oxygen planet. Why is com-bustion normally referred to as combination with oxy-gen? Why did we discuss aliens breathing oxygen, and exhaling carbon dioxide?

Only because we live on a planet in which free oxygen is a major component of the atmosphere. We do not think of this as unusual, but we ought to. As pointed out earlier, oxygen combines readily—even fiercely—with other elements. A planet with an oxygen atmosphere is unstable. If a world starts out with an atmosphere of pure oxygen, before long the normal processes of combustion will combine the oxygen to other, more stable compounds.

Clearly, that has not happened to the Earth. The presence of life, and in particular of plant life that performs photosynthesis, makes all the difference. Using the energy from sunlight, a plant reverses the process of combustion. It takes carbon dioxide and water, producing from them hydrocarbons, and releasing pure oxygen into the atmosphere. This is a dynamic, self-adjusting process. If there is more carbon dioxide in the air, plant activity will increase, serving to remove carbon dioxide and increase oxygen. Too little carbon dioxide, and plant growth decreases.

Because we grew up with this process, we tend not to realize how extraordinary it is. But the first life on Earth had to deal with an atmosphere containing no oxygen, but plenty of hydrogen. When the first photo-synthetic organism (almost certainly, some form of cyanobacteria) developed, a huge but unchronicled battle took place. To hydrogen-tolerant life, free oxygen is a caustic and poisonous gas. To oxygen-tolerant life, free hydrogen is an explosive.

The oxygen-producers and oxygen-breathers won, to become oaks and marigolds and tigers and humans. The hydrogen lovers remain as single-celled organisms, the anaerobic bacteria.

Free oxygen is so much a hallmark of life, James Lovelock (Lovelock, 1979) insists that the detection of substantial amounts of oxygen in a planetary atmosphere would prove, beyond doubt, that life must be present there. The converse, as shown by the early history of Earth, is not true: absence of oxygen does not mean absence of life. The science fiction writer is free to suppose that life has developed on other worlds in an atmosphere of hydrogen, or oxygen, or methane, or nitrogen, or carbon dioxide, or many other gases. Combinations are permitted, as our own atmosphere shows. But if you take the route of an exotic atmosphere, the chemical consequences must be worked out in detail. No atmosphere, please, of mixed oxygen and hydrogen.

The master of the design of alien planets and biospheres is Hal Clement. If you want to see how carefully and lovingly it can be done, read his *Mission of Gravity* (1953—with hydrogen-breathing natives, no less); *Cycle of Fire* (1957); *Close to Critical* (1964); and *Iceworld* (1953). If you want to see fascinating and exotic worlds that won't stand up to such close scrutiny, consult Larry Niven's *Ringworld* (1970), or wander the wild variety of planets to be found in his multiple volumes known collectively as Tales of Known Space.

TABLE 5.1
Materials, potential strengths

Element pairs*	Molecular weight (kcal/mole)	Bond strength	Strength to weight ratio
Silicon/carbon	40	104	2.60
Carbon/carbon	24	145	6.04
Fluorine/hydrogen	20	136	6.80
Boron/hydrogen	11	81	7.36
Carbon/oxygen	28	257	9.18
Hydrogen/hydrogen	2	104	52.0
Muonium/muonium	2.22	1,528	9,679
Positronium/positronium	1/919	104	95,576

* Not all these elements exist as stable molecules.

CHAPTER 6
The Limits of Biology

6.1 The miracle molecule. You will read in many places that if the twentieth century was from the scientific point of view the century of physics, then the one after it will be the century of biology. That should make the frontiers of biology of special interest to a science fiction writer. The question then is, where do we begin?

Fifty years ago, a writer on the limits of biology might have had trouble deciding where to start. The biological world offers such a dazzling diversity of forms and creatures at every scale, everything from bacteria and viruses to mushrooms and elephants. Today, there is no such problem. We have to begin with a single molecule, an organic compound with a long name but a famous abbreviation.

Deoxyribonucleic acid, universally shortened to DNA, was discovered in 1869 by the German chemist Friedrich Miescher. It was (and is) found in the nuclei of the cells of most living things, but no one knew its structure, what it did, or how important it was.

DNA is one of a class of chemicals known as *nucleic acids*. By the beginning of the twentieth century, the components of the DNA molecule were known to be sugars, phosphates, and two types of two chemical bases known as *purines* and *pyrimidines*. The functions of the molecule were still obscure, though in 1884 a zoologist, Hertwig, had written that it was the way that hereditary characteristics were passed on from generation to generation.

He was right, but most people didn't accept what he said. So when, in 1943, Erwin Schrödinger gave lectures in Ireland on the mechanisms of heredity, he did not talk about DNA. He proposed, in his lectures and in a short and very readable book *What Is Life?* (Schrödinger, 1944), that the basis for heredity must be some kind of code, in which specific sequences of chemicals were written and interpreted; however, he assumed that the "code-script," as he called it, was contained in proteins, in the form of an aperiodic crystal.

Schrödinger was right, in that heredity, and all cell reproduction, depends on what we now term the *genetic code*. But it took another decade before the nature of the code and the structure of the code carrier were determined.

DNA, not proteins, carries the genetic code, for humans and for everything remotely like us. Nature is prodigal with DNA. In most (but not all; mature red blood cells lack a nucleus) of the 10^{14} cells of our bodies, we have the DNA to provide a complete description of the whole organism. Your DNA is in all important respects the same as the DNA in any other animal or plant, everything from a wisteria to a walrus. The same, that is, in all important respects but one: your DNA defines the unique you, the walrus DNA defines the complete walrus. In principle, given one cell from my body a full copy of me could be grown. This idea of "clones" has been widely used in fiction (Varley, 1977, 1979, 1980), with some of the fiction posing as fact (Rorvik, 1978). Sheep and other mammals have been cloned, but no one has yet cloned a human. We can look for that in less than twenty years, regardless of laws passed by those who disapprove of the concept on religious or ethical grounds.

The structure of the DNA molecule was determined by Crick and Watson in 1953. The story of their discovery is told in frank detail by Watson in his book *The Double Helix* (Watson, 1968). The title is appropriate, because the molecule has the form of a double helical spiral. Strung out along the spiral, at regular intervals, are molecule after molecule of the four chemical bases. Their names are *adenine, cytosine, thymine,* and *guanine,* and they are exactly paired. Wherever on one strand of the double spiral you find a cytosine nucleotide base, paired with it on the other strand you will find guanine; if there is thymine on one strand, on the corresponding site of the other strand there will be adenine. If we were to read off the sequence of nucleotide bases along a single strand, we would find a long string of letters, A-G-T-G-C-T-A-A-C-C-G-T-A- (we are using the obvious abbreviations). The corresponding sites on the other strand would then, without a choice, read T-C-A-C-G-A-T-T-G-G-C-A-T-.

Long strands of DNA nucleotide bases, each base with an accompanying sugar and phosphate molecule, make up the *chromosomes* found in the nucleus of every cell. Individual genes, with which the science of genetics is mainly concerned, are subunits within the chromosomes. The division of the DNA into many separate chromosomes (humans have forty-six of them) seems to be mainly a matter of packing convenience. Efficient packing is necessary. There are about three billion separate nucleotide bases in human DNA, tucked into a cell nucleus only a few micrometers across. The chromosomes that define your body and brain (though not its contents—writers of cloning stories beware) are invisible to the naked eye.

As an interesting sidebar to the development of life, not all the DNA in a cell of your body will be found in the nucleus. Some is located in other small units, known as *mitochondria,* that control cell energy production. However, the DNA in mitochondria is not *your* DNA. It belongs to the mitochondria themselves, and it is used to control their own reproduction. It seems that the mitochondria were originally independent organisms, but long, long ago they abandoned that independence in favor of a symbiotic relationship with other creatures.

The means by which the DNA molecule reproduces itself is simple and elegant: the double helix *unwinds.* One branch of the spiral goes in one direction, the other in the opposite direction. As each site on the helix is left with an unpaired base, the correct pairing, cytosine/guanine or adenine/thymine, takes place automatically (the pairs of bases have a natural chemical affinity). The correct base is collected from a pool of materials within the cell. At the same time, the necessary sugar and phosphates are added to the spine of the helix. When the double helix has finished unwinding, *two* new double helices, each identical to the original one, have miraculously appeared. The DNA molecule has reproduced itself.

This copying procedure is incredibly accurate, assisted by a "proofreading" enzyme called DNA polymerase that can correct mistakes. Very rarely, however, there will be a glitch, perhaps an A where the copy ought to have a C, or a short sequence duplicated or left out completely. If this occurs in the reproductive cells of an organism, the copying error will pass on to the offspring. A change in DNA due to imperfect copying, or accidental damage, gives rise to what we term a *mutation*. Most mutations have no apparent effect, and of those that do, most lead to changes for the worse. The offspring, if it survives at all, will be unable to perform as well as the parent. Occasionally, however, there will be a favorable mutation. The new version will be an improvement over the original, and produce its own superior offspring. This is the driving mechanism of evolution.

However, we—you and I and the potted begonia in the corner—are not composed entirely of DNA. We have seen how the DNA molecule takes care of itself, but what makes the rest of us?

That calls for some agent to interpret and use the code contained in the DNA molecule. It is a two-stage process. First, another nucleic acid, RNA (ribonucleic acid), copies the information in the DNA molecule onto itself. It has a slightly different chemical composition (*uracil*, abbreviated to U, takes the place of thymine), but essentially it mimics the relevant DNA structure with matching bases. Note that, because RNA can match any sequence of sites in a DNA molecule, RNA can carry the same information as DNA.

What RNA cannot do, because it lacks the double helix

structure, is *make a copy of itself*. We will return to that later.

RNA copies information from the DNA molecule. Then it goes off to a place in the living cell where small round objects known as *ribosomes* are located. There, the RNA dictates the production of substances known as *amino acids*. The amino acids are the small and simple elements from which large and complex protein molecules are made. Just as DNA and RNA have strings of nucleotide bases, proteins have strings of amino acids. Each triplet of symbols in the RNA bases (A, C, G, and U) leads to the production of a unique amino acid. For example, the sequence U-A-C always, without exception, leads to the production of the amino acid tyrosine. The order of the sequence is important; C-A-U leads not to tyrosine, but to another amino acid, histidine.

Although each triplet leads to a unique amino acid, the converse is not true. There are 64 possible three-letter combinations, but they lead to only 20 amino acids. For example, both C-A-U and C-A-C produce histidine.

Now we have the final step: amino acids, created in the order dictated by the triplets (known as *codons*) in the RNA, in turn produce the proteins.

Interestingly, not all the three billion nucleotide bases in human DNA are used to make proteins. Only about ten percent of them do that. The other ninety percent, stretches of DNA that are known as "introns," don't seem to do anything at all. That may reflect current ignorance, and we will later learn what this "junk DNA" actually does. Simple organisms don't have introns—all their DNA is used to define the making of proteins. So why do more complex organisms have them?

Feel free to make up your own reasons. No one is in a position to disagree with you.

We have rendered down a century of work to a few hundred words, but the central message, stated by Francis Crick as the "Central Dogma of molecular biology," is simple: *DNA codes for the production of proteins; the process never, ever, goes the other way.*

There is another way of looking at this, and one that may sound more familiar. DNA controls reproduction, and also the production of proteins and hence our bodies. Nothing that we do to our bodies can ever go back and affect the DNA. In other words, there can be no inheritance of acquired characteristics.

6.2 The mystery of sex. Before we move on to other mysteries of biology, we need to answer an implied question. We can be regarded, as Richard Dawkins has eloquently pointed out in *The Selfish Gene* (Dawkins, 1976), as nothing more than large-scale mechanisms designed to propagate our own genetic material. To most organisms, DNA is the most precious thing in the world, the only way to assure that their line continues. Few people would argue, seeing the powerful imperative to propagate as we see it displayed throughout the living world, that the preservation and multiplication of genetic material is the Prime Directive of nature.

However, when we examine the subject of sex logically (a mental exercise for which, as any newspaper will show, humans show little apparent aptitude) we find a paradox. Your DNA is high-quality stuff, developed and fine-tuned over four billion years. It is you, the essence of you, the only way for you to continue an existence in the future (let us leave aside for the moment the notion of your and my possibly immortal prose).

So what do you do? You mate, with a genetic stranger. At that point your unique and wonderful DNA becomes mixed, fifty-fifty, with other DNA about which you know very little. Even if you have known your partner all your life, it is still true to say that the two of you are strangers at the DNA level. Indeed, the best bet from the point of view of your genes would be to mate with a close relative, where you share a high proportion of common genetic material.

This is not, of course, what happens. Incest is taboo in most human societies and mating outside the family seems generally preferred everywhere.

What is going on? Why, taking a gene's-eye view of things, is sexual reproduction such a big hit? Why do all the most complex life forms on Earth employ, all or part of the time, sex as a tool for propagation?

I do not think that biology today offers complete answers to these question. Richard Dawkins at one point seems all ready to tackle it in *Climbing Mount Improbable* (Dawkins, 1996). But then he veers away, or at least postpones: "But then the whole question of sex and why it is there . . . is another story and a difficult one to tell. Maybe one day I'll summon up the courage to tackle it in full and write a whole book about the evolution of sex."

I wish he would. Meanwhile, here is a brief analysis, some of it based on personal speculation. In summary, the main idea is that the driving force for rapid change, and hence for exploiting a changing environment, is *selection*, not mutation.

Let me repeat and rephrase an earlier statement which I believe is not controversial: changes that take place in an organism over time, as a result of random mutation, take place *slowly*. Each mutation may be harmful, beneficial, or neutral in terms of survival of the organism's offspring. Beneficial mutations will prosper (there is something of a tautology here, since the definition of *beneficial* is that organisms with the mutation do well). However, significant changes as a result of mutation require many thousands of generations.

Also changing over time is the environment in which the organism lives. The environmental changes may be slow (tectonic forces that raise mountain ranges) or fast (earthquake and volcano), but in any case their rates of change are largely independent of the rates of organism mutation. I say largely, because changes in chemistry or radioactivity levels certainly affect mutation rates.

Consider changes in environment which take place over time scales that are, in terms of mutation rates, very fast. A volcanic eruption, like Mount Pinatubo in 1993, fills the upper atmosphere with dust and cools the atmosphere by a few degrees. El Niño, in 1998, causes anomalous heating of the seas. A large calving of the Antarctic ice shelf reduces salinity over much of the southern oceans.

An organism which reproduces asexually will adapt to such changes to the limits of its variability. We can use the term "natural selection" to describe this process, but it will not normally be mutation. An organism cannot mutate fast enough to be useful, nor can it modify its own genetic material. It passes on to the next generation an identical copy of what it possesses.

Now consider sexual reproduction. The mixing of genetic material permits a great variety in offspring, in both appearance and function. Thus adaptation of organisms, accompanied by rapid morphological changes, can be far faster than mutation would permit. Consider the "unnatural selection" process that has led to forms of a single species as disparate in size and shape as the Chihuahua and the Great Dane, during the relatively short period of human domestication of animals. Morphological evolution can be *fast*, when something (humans or Nature) drives it. It will be slow when there is no driving force for change. However, in either case the available pool of DNA for the whole sexually-compatible group of organisms is unchanged, though grossly variable at the individual level. Thus a species can adapt and thrive, using sexual reproduction, without waiting for the slow process of favorable mutation. This is a huge evolutionary advantage.

We still have to address an important question: Is it possible for changes to take place in an organism, sufficient that we can say we now have a new species, other than by the slow processes of mutation? If not, then although in the short run sexual reproduction has an advantage over asexual reproduction, in the long term that advantage diminishes.

I argue that there is also the following long-term gain in sexual reproduction. When a male and female produce offspring, they mix their DNA fifty-fifty. However, this is not a random mixing. Certain very specific segments of DNA, which we call genes, come from one parent, and this is an all-or-nothing process. Thus, an offspring gets that whole segment from one parent, or from the other. It does not get half and half, or if it does get a fractional gene, the result cannot survive. Since there are thousands of segments (genes) we have a gigantic number of possible offspring, with all sorts of gene mixes.

Now take one group of offspring away to a different

environment. Natural selection takes place, and some gene segments, rather than existing in the population equally in their two possible forms, are preferred because of environmental pressures in just one form. Offspring with that form thrive, others fail. The organisms begin to look and act differently from the original form, because their gene choices are selected to suit the new environment. Finally, one form of a gene may exist in the new environment, while the complementary form of the gene, selected against, does not. It has been removed completely from the organisms in that environment.

In the same way, in some other environment, other genes have a preferred form for organism survival. Their complementary form does not exist in that environment.

If mutations did not occur and we put specimens of organisms from the two environments back into their original settings, they would mate and their offspring would have all the original forms of genes.

However, we cannot ignore mutation completely. It is a random process, but it happens. A beneficial mutation will spread rapidly through a population. We might say that we had a new species every time such a mutation occurred and spread, except that we will normally have no way of observing such a change. Over time, however, there will be recognizable changes, and we then say that the organism has evolved. We would see the evolution of a single species, whether or not the organism propagates sexually.

Now here is the key, if obvious, piece of the argument: *mutations cannot occur in genes that are not present in an organism.* Different environments, for sexually reproducing organisms, will have different mutations. At some point, the original organisms that were placed into two different environments will be different not only morphologically in appearance and behavior (accomplished via sexual selection of genes), but through the accumulation of different changes in their actual genetic make-up. The new versions of genes will not be compatible with the old complementary set of genes. We see speciation. One species has become two. And that process, the creation of new permanent forms, is easier and faster with the aid of sexual reproduction. Sex is, in fact, a good thing.

At this point, I ought to say that not everyone agrees as to *why* sex is a good thing. Steven Pinker, in *How the Mind Works* (1997) supports a different theory as to why sex was a valuable invention for living creatures. First, he points out that an organism cannot practice any policy that implies present sacrifice for future benefit. "Playing on the come" will not work, since everything from squash to squids must maximize the number of its *immediate* descendants. (Not only that, an organism does not sacrifice itself, even for the good of the species, unless there are sound reasons, based on the selfishness of genes, for doing so. This has caused workers, including Dawkins, considerable trouble, explaining how altruism can also be a form of self-interest.)

Pinker favors a theory proposed by John Tooby, which claims that sex was developed as a way of protecting organisms against disease. The argument goes as follows: We are invaded all the time by a variety of tiny critters, who see us as a plentiful food supply. We have built up protections against them, but they in turn have become very cunning at penetrating our defenses. When an organism employs asexual reproduction, and some parasite organism finds a way around the defense, the game is over, because the same trick will penetrate the defenses of future generations with identical genetic make-up. Sexual reproduction, however, scrambles the genes, and makes the offspring less susceptible to parasitic invasion. Thus, sex provides a partial fresh start with each generation.

I am less persuaded than Pinker by this argument (although I strongly recommend *How the Mind Works* for a hundred other good reasons). It seems to me that there is no inconsistency between optimizing for the present generation, and sexual reproduction. In fact, the mixing of genes that sex offers increases the total variation in the next generation, without the dangers presented by mutation (which is normally *unfavorable* to an organism), and therefore improves the short-term odds.

Which theory is right? I don't know. Nor, I argue, does anyone else. However, this is not an either/or situation, where one theory must be right at the expense of another. Perhaps sexual reproduction allows organisms to adapt more rapidly to new environmental niches, and also serves as a defense against disease.

Is there a third reason why sexual reproduction has been such an overwhelming success? Feel free to conjecture. Alternate scientific theories are exactly the place where science fiction stories flourish. And if you would like to read a radically different suggestion as to why evolution seems to proceed far more rapidly than simple mutation would suggest, try *Paths to Otherwhere* (Hogan, 1996), where the subject is dealt with, amazingly, in terms of the many-worlds theory of Everett and Wheeler (see Chapter 2).

6.3 Viruses, RNA, prions, and the origin of life. The story of DNA seems astonishingly simple and complete. Let us ask the usual questions: What don't we know, and what do we know what ain't so?

For one thing, we don't know how this whole process started.

The interdependence of the proteins and the DNA is a highly improbable connection. To make a new cell, both are needed. If you lack either one, the process cannot work. It seems ridiculous to suggest that both DNA and the necessary protein production factory could have developed independently of each other, and work together without a hitch. It is as though you developed a car body while I, without consulting you, developed an engine, neither of us having done anything like it before. We put them together, and the whole automobile runs like a dream.

It would be a dream. That independent development of DNA and proteins is obviously not what happened. But what did?

To provide a possible explanation, we go to the world of viruses. At first that may seem to make the problem worse. A virus is a mystery organism (but a godsend, I sometimes feel, for the medical profession. The doctor's pronouncement, "You have a virus," is often the equivalent of, "I don't know what is wrong with you, but I know I can't give you anything to cure it.").

Viruses are minute, much smaller even than cells. Their small size is possible because they lack a cell structure or a protective cell wall, and they don't have their own ribosome protein factories. All they are is a tiny chunk of DNA, wrapped in a coat of protein. Some of them also have little tails.

It is possible that viruses are degenerate forms, organisms that once possessed the full machinery for self-reproduction but at some point abandoned it. Be that as it may, we must still explain how something so small, on the very borderline between living and nonliving, can go about reproducing itself when it has none of the equipment we have described as necessary. If we find the answer to that, maybe we will solve the problem of the separate development of proteins and DNA.

A memorable report in a British newspaper of a divorce court proceeding a few years ago ended as follows (with minor changes as to names): "Living at the time as a paying house-guest of Mr. and Mrs. Smith was Mr. Jones, a man with an artificial leg. One day Mrs. Smith asked her husband, if a woman had a baby by Mr. Jones, would the baby have an artificial leg? Mr. Smith then began to be suspicious."

If, metaphorically speaking, the paying guest in your house happened to be a virus, then the chance of your cells having an artificial leg would be very good indeed.

What happens is this. The virus penetrates the wall of a normal, healthy cell, often with the aid of its little tail of protein. Once inside, the virus takes over the cell's own copying equipment, using it to produce hundreds or thousands of new viruses until the chemical supplies of the cell are used up. Then the cell wall bursts, releasing the viruses, which go on to repeat the process. The virus doesn't carry its own protein factory, because it doesn't need it. Viruses are, and must be, parasitic on cells.

Again, the story seems neat and complete, but not useful to resolve our mystery of how the whole process began. Then, to add confusion, certain viruses were discovered that have no DNA at all.

What they have is RNA. Such viruses are known as *retroviruses*, and they are famous, or infamous, because their number includes the Human Immunodeficiency Virus, HIV, associated with the disease AIDS. (The naming of the HIV virus, and the battle over priority of discovery, is an astonishing story that I won't go into here. Science is the search for absolute truth, and scientists are objective, dispassionate people. Right? Look out of the window, and you will see the Easter Bunny.)

How can something without DNA reproduce? We have emphasized the importance of the DNA double spiral, which RNA lacks. A retrovirus has to work hard indeed to produce the next generation. First, it invades a cell. Next, it uses the one-to-one correspondence between its own RNA bases (A, C, G, U) to make matching DNA (T, G, C, A). Then it employs the cell's own DNA-reproducing mechanism to make DNA copies. Finally, the virus employs the rest of the cell machinery to make matching RNA (its own genetic material) and hence more copies of itself.

Here, then, is an interesting possibility. Since some organisms have no DNA, but do have RNA, suppose that RNA came *first*, and DNA was a later development? We know that RNA can produce proteins, and it doesn't need DNA for that. This central early role for RNA has strong advocates, particularly since RNA has been found to contain *ribozymes*. Not to be confused with ribo*somes*, ribozymes are enzymes able to snip and reorganize the sequence of nucleotide bases in the RNA itself.

The argument for RNA is interesting, but not yet persuasive. We can make proteins, yes, but without the DNA double helix for exact copying, RNA, with or without the protein factory, cannot produce the next generation.

Where might we find a method of reproduction that does not need DNA, but might lead to DNA and RNA's eventual development?

The answer, even twenty years ago, would have belonged to Chapter 13. Today, thanks in large part to Stanley Prusiner's receipt of the 1997 Nobel Prize for

Medicine, the idea is in the scientific mainstream. However, it was pure scientific heresy when Prusiner, in the late 1970s, decided that what we "know" in molecular biology is possibly not so.

What we know is the Central Dogma, according to which DNA, working via RNA, produces proteins. Proteins do not reproduce, nor can their actions affect the DNA. As an organism succeeds or fails in the world, so will its DNA be more or less present in the world.

Prusiner had been studying certain peculiar diseases with a long incubation time. They include *scrapie*, mostly affecting sheep and goats; *kuru*, the "laughing death" disease of the natives of New Guinea, that became famous because cannibalism was involved in its spread; *Creutzfeldt-Jakob* disease, a rare and fatal form of dementia and loss of coordination in humans; and, most recently, the "mad cow disease" (*bovine spongiform encephalopathy*) that required the killing by farmers of millions of cattle in Great Britain. These diseases have a long latency period before the infected animal or human shows symptoms, and the standard theory was that a "slow virus" was responsible for them.

If that were the case, the infecting agent for the diseases would have to contain DNA, or, if this happened to be a retrovirus, RNA. However, the analysis of infecting material showed no evidence of either nucleic acid. Finally, in 1982, Prusiner proposed that the infectious agent for scrapie and related diseases consists exclusively of protein. The term *prion* (pronounced pree-on), for "proteinaceous infectious particles," was introduced. Soon afterwards, Prusiner and his co-workers discovered a protein that seemed to be present always in the infectious agent for scrapie. They termed it PrP, for "prion protein." Moreover, the same protein occurs naturally, in animals that are not sick. There is a tiny chemical difference between the two forms of PrP, amounting to a single amino acid. The bigger difference, however, is not chemical but *conformational*. In other words, PrP can exist in two different molecular shapes.

This suggests a mechanism by which infection can take

place. Call one form of the protein PrP^i, i for infectious. Call the other PrP^n, n for normal, since it is normally present in the body. When PrP^i invades the body, it induces the PrP^n molecules that it encounters to change their shape to its shape. It may also force the substitution of a single amino acid. The changed forms become PrP^i molecules, which can then go on and modify more PrP^n. Eventually, there are so many PrP^i molecules that the body begins to display symptoms of the disease.

We have found, with prions, a way of reproducing PrP^i without the use of DNA. All we needed was a supply of PrP^n.

We seem to have run far afield from our original question, of how the process of reproduction started, but we may in fact be close to answering one of the most basic questions of all: What was the origin of life?

Given a supply of energy and basic inorganic components, it is not difficult to produce amino acids. That was shown experimentally by Stanley Miller, back in the early 1950s. Further, if we had a plentiful supply of various kinds of amino acids, floating free in the early seas of Earth, they would naturally combine to produce a variety of different proteins. And if one of those, an ur-protein that was some ancient relative of PrP^i, could induce a conformational change and a minor chemical change in other proteins, so that they became exact copies of itself, we would have *reproduction*. The spread of our ur-protein would be limited by the extent of its "food supply"; i.e., the other proteins in the ancient ocean. However, evolution would be at work, creating modified, and sometimes improved, ur-protein.

It is easy to suggest a direction of improvement. The ur-protein would be more successful if the range of other proteins that it could adapt to its own form could be increased. One way to do this would be through the assembly of the necessary protein from simpler, smaller units—ideally, from the basic amino acids themselves.

This is very close to the processes of eating, digestion, and reproduction so familiar to us today. RNA and DNA could *evolve* from ur-proteins, as more efficient methods of reproduction. The best process of reproduction would naturally employ the simplest, and therefore the most widely available, components.

Is this what actually happened, in life's earliest history?

No one knows. In fact, no one has any mechanism to offer more plausible than the one offered here. As fiction writers, we can go with a prion approach; or we can assume any other that leads in a rational way to today's world of living creatures.

6.4 Local, or universal? Life elsewhere. It is possible that life did not begin on Earth, but was brought here. That does not avoid the question of an origin—we are then forced to ask how that life came into being—but it does remove a constraint on the speed with which life had to develop. We know that life was present relatively early in Earth's history. The planet is about four and a half billion years old, and there were living organisms here close to four billion years ago. Did they develop here, or were they imported from outside?

That suggests another question: Did primitive Earth possess the warm, placid oceans often pictured in considering the origin of life (we have hinted at it already as the womb for our ur-protein); or was the early world all storm and violence?

Modern ideas of solar system formation and evolution favor the latter notion. The early system was dominated by celestial collisions on the largest scale. Great chunks of matter, some as big as the Moon, hit Earth and the other planets of the inner system. The impact effects must have been prodigious. For example, a small asteroid one kilometer in radius would release into the Earth's biosphere four hundred times as much energy as the biggest volcanic eruption of modern times (the 1815 explosion of Tambora in Indonesia; the following year, 1816, was known as "the year without a summer" because crops failed to ripen). The asteroid whose arrival is believed to have led to the extinction of the dinosaurs (see Chapter 13) was a good deal bigger, perhaps ten kilometers across, and it delivered a blow with 50,000 times the energy of Tambora's eruption. Even so, it is minute in size compared with many of the bodies roaming the solar system four billion years ago. Our Moon, for example, has a diameter of about 3,500 kilometers.

Earth in its early days was subject to a deadly rain from heaven, each impact delivering the energy equivalent to hundreds of thousands of full-scale nuclear wars. As long as this was going on, it would surely be impossible for life to develop.

Or would it? Not if the idea proposed by A.G. Cairns-Smith in several books, (Cairns-Smith, 1971, 1982, 1985) turns out to be true. He suggests that the site of life's origin was not some warm, amniotic ocean, but that the first living organisms formed in *clays*. This life was not based on DNA and RNA. Those were later developments, taking over because they were more efficient or more stable. The original self-replicating entities were inorganic crystals. Clays are a perfect site for such crystals to form, because clay is "sticky," not only in the usual sense of the word, but as a place where chemical ions readily attach and remain. Life could begin and thrive in clays at a time when Earth was still too chaotic and inhospitable for anything like today's organisms to survive.

The Cairns-Smith idea does two things. It offers another possibility for the beginnings of life on Earth; and it makes us wonder, under what strange conditions might life arise and thrive? We know of organisms three miles down in the ocean, living at pressures of two hundred atmospheres near vents of superheated water, and dying if the temperature drops too far below boiling point. Many of these creatures depend not on photosynthesis for their basic energy supply, but on sulfur-based chemosynthesis.

Could anything sound more alien?

It could. For one thing, all these organisms are just like us, in that they employ DNA or RNA in their genetic codes. Recently, a full genome (the complete code sequence) was developed for a deep ocean bacterium known as *Methanococcus jannaschii*. Its 1.66-million base bacterial genome was built up from the A, C, G, and T nucleotides. Currently, a huge effort goes on to provide the complete mapping of the human genome, with its three billion nucleotide base pairs.

We have no idea if life must be this way, right across the universe. Is an intelligent mud possible, a crystalline matrix layered like the silicon chips of our computers? Or on the massive planet of an unnamed star in the Andromeda Galaxy, do the creatures wriggling across the seabed inevitably possess a molecular pattern, G-C-T-A-A-G-, that we would recognize at once?

Are there a million different ways of writing the book of life, only one volume of which we have seen? Or do they all have a double helix? Why not a triple helix? That would permit better mixing of genetic material during sexual reproduction, with one-third provided by each of the three parents.

A few years ago, these questions seemed unanswerable. The recent discovery of possible archaic life in a meteorite that originated on Mars has changed everything. The Mars Sampler spacecraft, visiting the planet and examining the surface, is currently scheduled for 2005, but may be advanced to 2003. In six or seven years, we may know: are DNA and RNA local to Earth, or do they represent a more universal solution?

6.5 Aging and immortality. Juan Ponce de Leon searched for years, but he did not find the elixir of life and the secret of perpetual youth.

Perhaps he was looking in the wrong place. Just as the study of the faint astronomical objects known as galaxies led us in Chapter 4 to the age of the universe, the study of the curious double helix within our own cells may lead to the understanding of aging, and its ultimate reversal.

I believe that this understanding will come in a decade or less. Would-be storytellers should start now, or be too late.

Chromosomes are made of long strands of DNA. At its very end, each chromosome has something called a *telomere*. This is just a repeated sequence of a particular set of nucleotide bases. The sequence is not used for RNA or protein production, and for most organisms the repeated set is mainly thymine and guanine. Human telomeres, and those of all vertebrates, are T-T-A-G-G-G, repeated a couple of thousand times; roundworms have T-T-A-G-G-C. At the most basic level, we are not much different from worms.

The telomeres serve a well-defined and useful purpose. They prevent chromosomes sticking to each other or mixing with each other, and hence they aid in stable and accurate DNA replication. However, the telomeres themselves are not stable bodies. They repeatedly shorten and (in certain cases) lengthen.

For example, when a cell divides and DNA is copied, the copying does not extend right to the end of the chromosome. A small piece of the telomere is lost. Over time, if no compensating mechanism were at work, the telomere would disappear. The chromosomes would then develop the equivalent of split ends, and vital genetic information would be lost.

This does not happen, because an enzyme called *telomerase* generates new copies of the telomere base sequence and adds them to the ends of the chromosomes. The telomeres will then be always of approximately the same length.

The presence of telomerase in single-cell organisms allows them to be effectively immortal. (Some bacteria have no need of telomores and telomerase. Their DNA is arranged in the form of a continuous ring, and it can therefore be completely copied.) They can divide an indefinitely large number of times, with the vital DNA of their genetic code protected by the telomeres. However, many human cells are devoid of telomerase. As has been known since the work of Leonard Hayflick in the 1960s, human body cells are able to reproduce only a certain number of times; after that they become, in Hayflick's term, "senescent" and eventually die. Moreover, cells from a human newborn can divide 80 to 90 times when they are grown in a suitable cell culture; but cells from a 70-year-old will divide successfully only 20 to 30 times.

In the 1970s, an explanation was proposed. Without telomerase, the chromosomes lose part of the telomere at each cell division. Eventually, there are no protective telomeres left. Cell division ceases, and the cell dies. This also provides at least a partial explanation of human aging. If cells are not able to keep dividing, body functions will be impaired.

There is still a mystery to be explained. Although normal human cells die after a limited number of divisions, the same is not true of cancer cells. They will go on growing and dividing in culture, apparently indefinitely. Not surprisingly, in view of what we have learned so far, cancer cells produce telomerase. They do so, even when they derive from body cell types in which telomerase is absent.

We now see two exciting possibilities. On the one hand, if we could prevent the production of telomerase we would inhibit the spread of cancer cells, while not affecting normal cells which already lack telomerase. On the other hand, if we could stimulate the production of telomerase in all the cells of our bodies, cell division would not result in the gradual destruction of the telomeres. Tissue repair would take place in the 70-year-old at the same rate as in the newborn. The aging process would be halted, and perhaps even reversed.

There is a fine balance here, one which we do not yet know how to maintain. Too much telomerase, and the cells run wild and become cancerous (though there is evidence, based on the short telomeres of cancerous cells, that telomerase is produced only after a cell begins to multiply uncontrollably). Too little telomerase, and the aging process sets in at the cellular level. After a while the effects are felt through the whole organism.

There is one other factor to note, and it suggests that we do not have the full story. During the duplication of DNA associated with the reproduction of a multi-celled organism, as opposed to cell division within the organism, the telomeres are somehow kept intact. This is absolutely essential, otherwise a baby would be born senescent. But how is the body able to preserve the telomeres in one type of copying, while they are degraded in another?

There is one simple explanation, though it may be a personally unpalatable one. Aging and death are desirable from an evolutionary point of view. Only by reproducing do we open the door for the biological improvements that keep us competitive with the rest of Nature. Only by mortality do we provide the best assurance for the long-term survival of our DNA.

This runs counter to our desire for personal survival, but in the words of Richard Dawkins, "DNA neither knows nor cares. DNA just is. And we dance to its music" (*River Out of Eden*, Dawkins, 1995). That has apparently been true for all of life's long history. But when we fully and finally understand telomeres and cell division, perhaps we will have a chance to dance to music of our own choosing.

Telomeres, and their role in aging and the prevention of aging, have already been used in science fiction. See, for example, Bruce Sterling's *Holy Fire* (1996) and my own *Aftermath* (1998).

6.6 Aging: a second look. We have proposed the erosion of the telomeres as a mechanism for aging, but it is unlikely to be the only factor. For one thing, people usually die long before they reach the "telomere limit" at which chromosome copying is impaired. This implies reasons for aging that go beyond what is happening at the level of the single cell. If we again invoke Philip Anderson's "More is different" argument, the large, complex assembly of the human body has properties that cannot be deduced by analysis of its separate components. Consciousness, to take one example, does not seem to exist at the cellular level. It emerges only when a sufficiently large aggregate of cells has been created.

Humans have uniquely large brains as a fraction of total body mass. We do not know if we are the only creatures on Earth with consciousness of self, but we do know other, well-measured ways in which we are unique. For instance, we have a peculiarly long life span for animals. The sturgeon (150 years) and the tortoise (140 years) definitely live longer, but among mammals only those land and sea giants, the elephant (70) and the whale (80), come close to human maximum life expectancy. The difference between humans and other animals is more striking if we work in terms of number of heartbeats. With that measure, we live longer than anything. Big brains seem to help, though we don't know how.

I am taking it for granted that we would all like to live longer, provided that we can do so in good health. Aging and death may be necessary from an evolutionary point of view, but from a personal point of view both are most undesirable. If we cannot escape death, can we at least postpone aging?

At the cellular level, one class of frequently-named suspects as causes of both aging and cancer is free radicals. Vitamin C (discussed further in Chapter 13) and Vitamin E neutralize free radicals.

Whether or not the secret of human longevity involves the whole organism rather than being defined at the single cell level, we can certainly identify particular organic changes associated with aging. Two that continue to arouse a great deal of attention involve the thymus organ and the pineal gland.

The thymus is small. Less than half an ounce at birth, it sits above the heart and is about an ounce at maximum size, close to puberty. After puberty the thymus begins to shrink, and becomes inactive by age forty. It is an important part of T-cell production and hence of the body's immune response system.

The pineal gland is small also, about a centimeter across, and it sits at the base of the brain. Its main known function is the production of melatonin. The pineal gland begins to diminish in activity very early in life, with changes already occurring by the time we are seven or eight years old. Like the thymus, the pineal seems to close down its activity completely by the fortieth year.

The medical profession insists that the use of drugs to substitute for the output of the thymus organ and the pineal gland, or to stimulate their renewed activity, or even to reduce the number of free radicals in the body, is a total waste of time and money. The general public, on the other hand, often seems to agree with the student who answered the question on an English test, "What is a word for an ignorant pretender to medical knowledge?" with "A doctor."

At any rate, there is current widespread interest in such nonprescription drugs as melatonin, Co-enzyme Q, Vitamin C, Vitamin E, Vitamin B-6, Vitamin B-12, DHEA (de-hydro-epi-androsterone), and SAM (S-adenosyl-methionine). In less than ten years we will have evidence as to whether these diet additives have any beneficial effect on the aging process. Meanwhile, many people are not waiting. The drugs may be no more than stopgap measures, retarding aging but certainly not halting or reversing it; but, the logic goes, that's a good deal better than nothing.

One other whole-body function seems to correlate with aging. The first signs that we are beginning to age appear when our bodies stop growing. Moreover, animals such as carp, which grow continuously, also seem to live indefinitely long. They die only when some disease or predator disposes of them.

Continuous growth hardly appears an answer for humans. I doubt if anyone wants to be twelve feet tall and fifteen hundred pounds, unable to move or even to stand up. But might there be some "growth extract" that we could take from animals, to increase our own life expectancy?

I'm not optimistic. In any case, the science fiction story using that idea was written long ago. In Aldous Huxley's book *After Many a Summer Dies the Swan* (Huxley, 1939), an eccentric old oil magnate adopts the unpleasant diet of "triturated carp viscera"—chopped-up carp guts. He lives to be over two hundred years old, but at a price. Like Tithonus, who asked the gods only for immortality, he does not die but continues to age. As he does so he goes through a process of devolution, by the end of the story becoming an ape.

I'm not sure any of us want that. On the other hand, for another century or more of life . . . maybe just a trial taste?

6.7 Tissue engineering. It is a great annoyance when the "dumb beasts" of the animal world do things that we supposedly super-smart humans cannot; not just things based on specialization of body structure, such as flying like an eagle, swimming like a dolphin, or jumping like a flea, but things which by all logic our bodies should be able to manage without modification.

Why can't we hibernate or estivate, slowing our metabolism in times when food or water is short? Surely that was once a valuable survival mechanism, even if food for many of us is now almost too easily available. Still of importance today, why can't we grow a new finger or foot if we lose one, or connect a spinal cord severed by injury? We grow new skin without any problem, so some regeneration capability is clearly built into us. But amphibians can grow whole new limbs, which means they also have the capacity to regenerate nerve cells.

If we cannot regrow a limb or an organ, you might think that we ought at least to be able to accept one from some other human donor. The heart is nothing more than a pump, and one person's pancreas is in all important details exactly like another's. Livers, spleens, testicles or ovaries, hearts, wombs, lungs and kidneys are functionally identical in you and in me. It seems reasonable that you should be able to take one of my kidneys if yours are failing.

As the first surgeons to attempt organ transplants quickly learned, it's not so easy. The operation is relatively straightforward, but unless the donor happens to be your identical twin there is a big danger of organ rejection. The body treats the new part not as an essential and helpful component of itself, but as an intruder.

The problem lies not at the organ level, but at the cell level. Our bodies, as part of their defense mechanisms against invading organisms, seek out and destroy anything that does not carry the correct chemical markers that denote "self." The body functions that perform such recognition, and label something as "friend" (ignore) or "foe" (destroy), are known collectively as the *immune system*. Identical twins have the same immune system, and transplants between such twins are not rejected. Lacking an identical twin, your chances of a successful transplant are best if your organ donor is a close relative.

Today, organ transplants are usually accompanied by drugs that inhibit the action of the immune system. That, of course, carries its own risk. What happens when the bacteria of disease enter your body after a transplant operation? Without your immune system to recognize and devour the intruders, bacteria will multiply freely. You will die—not from organ rejection, but from some conventional infection.

Transplant patients live on the fine edge between two dangers. Too many immune system inhibitors, and infection gets you; too few, and the new organ is rejected by the body. When the immune system is weakened, it is vital to recognize the signs of disease and use antibiotics and other drugs to fight it in its earliest stages.

Is there a way out?

There is, but it is not yet a standard part of the medical community's arsenal. It is known as *tissue engineering*.

The basic idea is simple. Suppose that one of your body organs is failing. To be specific, let us suppose that it is your kidney. Even a diseased kidney has healthy cells. If we could just take a few of those cells, and encourage them to divide and multiply in the right way (including making structural components of the kidney, such as veins and arteries), then we could grow a whole kidney outside your body. When we performed the transplant, the new kidney would be in no danger of rejection. The immune system would identify the replacement organ as "self."

Unfortunately we cannot grow a kidney *in vitro*, using some nutrient bath; and if we try to grow a copy of one of your kidneys in some other person or animal, the host's immune system will send up the red flag that denotes "enemy," and proceed to destroy the intruder cells before they can begin the task of kidney construction. Again, we seem to be stymied.

However, occasionally an item appears in the news about a "bubble child." This is a person who has been born without a working immune system. The only way this unfortunate can survive is by complete isolation from all people and diseases. It is a precarious existence, and the fact that such a person could in principle accept any organ transplant without rejection is little consolation.

What nature occasionally does to humans, scientists have been able to do with animals. Lines of mice and rats have been bred that lack immune systems. They will not reject foreign tissue introduced into their bodies. Suppose that we introduce under the skin of such an animal a mold of porous, biodegradable polymer, configured to match the shape and structure of a kidney. We "seed" this mold with cells from your own kidney. These cells will be nourished by the blood of the host mouse or rat. They will multiply, to produce a whole kidney as the biodegradable "scaffolding" dissolves away. There will finally be a whole kidney, ready for removal and use as a replacement for your own failing kidney.

That is the idea. The execution, to make any organ we choose, is years in the future. At the moment there has been success only with the growth of cartilage. The other organs mentioned represent a far tougher problem.

There is also the problem that a mouse or rat is much too small to support the growth of a human liver weighing three pounds or more. In addition, some people would certainly find such a use of animals inhumane and unacceptable.

My own preferred solution to both problems is simple. The one living organism in the world whose immune system is guaranteed not to reject my tissue is me. When tissue engineering is perfected, I will grow copies of my own heart, lungs, and other necessary organs, on or in my own body, in advance of need. When full-grown they will be removed from me and placed in cold storage until the time comes to use them.

As a final note, let us recognize that for some diseases organ replacement will never be an option. This is the case with anything affecting the brain. Alone of all our organs, the brain contains our sense of identity. Another approach can then sometimes be used. Fetal tissue has not yet developed its own characteristic signature for immune system recognition. Thus, implanted fetal tissue is less likely to suffer rejection by the host body. Parkinson's disease is characterized by a loss of dopamine production. The implanting of fetal dopamine-producing tissue in a patient's brain alleviates the worst symptoms of the disease.

The most effective such tissue is human fetal tissue.

The treatment does not, of course, produce a cure. It also leads, in an aggravated form, to ethical questions similar to those arising whenever animals or humans become a part of human medical procedures.

A discussion of other ethical questions and possible societal response to tissue engineering can be found in a novel by Nancy Kress, *Maximum Light* (Kress, 1998).

CHAPTER 7
New Worlds for Old

The solar system has provided a wonderful, fertile field for speculation since the earliest days of science fiction. Set your stories there, by all means; but unless you want those stories to be dismissed as fantasy by the critical reader, make it the *new* solar system, as revealed by recent observations.

Even fifty years ago, the writer had lots of freedom. Telescopic observations of the Sun, Moon, and planets had told us a fair amount, but that was overwhelmed by the things we didn't know—what does the other side of the Moon, never seen from Earth, look like? What is beneath the perennial clouds of Venus?

Today, those and many other mysteries have gone away. Planetary probes have had a close-up look at every world except Pluto. Space-based telescopes have given us not only images, but spectroscopic data about all the planets.

We will confine this chapter to the "edges" of the solar system—not in terms of location, but in terms of knowledge. We will seek virgin territory for storytelling, where there is still hope for surprises.

7.1 Mercury. The planet closest to the Sun is Mercury. Before 1974, this was thought of as an airless ball, moving around the Sun in a rather elongated ellipse every 88 days. It was believed to present the same face to the Sun all the time, so that one side would be fiercely hot, and the other chillingly cold. Astronomers knew that Mercury had little or no atmosphere. A planet closer to the Sun than Earth sometimes passes between us and the Sun. Sunlight will then be refracted by any substantial atmosphere. There is no sign of that, so the surface of Mercury must be close to a perfect vacuum.

The big change in our knowledge of Mercury came with the Mariner 10 spacecraft, which in 1974–75 performed a series of flybys of the planet. It sent back pictures from three close encounters, and produced the first big surprise: the surface of Mercury looks at first sight exactly like the Moon. It is cratered, barren, and airless. Mariner also discovered a magnetic field, about one percent of Earth's. This, together with the planet's high density, suggests a substantial iron core maybe 1,500 kilometers in diameter. (Mercury itself is only 4,500 kms. in diameter.) At least part of that core should be fluid, allowing the existence of a permanent dynamo that generates the external magnetic field.

Mercury's rotation period was another surprise. The old assumption, that tidal forces would have locked it in position to present the same face to the Sun all the time, turned out to be wrong. If that were the case, the rotation period of Mercury would be the same as its year, 88 days. Mercury actually goes through one complete revolution on its axis in 58.6 Earth-days. It is no coincidence that 58.6 is two-thirds of 88. A dynamical effect known as a "resonance lock" keeps those two periods in that exact ratio. As one odd result, a *day* on Mercury lasts exactly two of its *years* (because the planet turns *one and a half times* on its axis in the time it takes to make one full circuit around the Sun). Since the planet does not present the same face to the Sun all the time, all sides get baked; the planet is hot all over, except possibly at the very poles, rather than just on one side as was previously thought.

Mercury has probably changed little in appearance in the past three billion years. However, it has one interesting difference from the Moon; its surface is more wrinkled, probably as a result of more cooling and contraction than the Moon has ever experienced. On the other hand, anything three billion years old has a right to be wrinkled.

The "old" Mercury allowed some fascinating science fiction stories to be written about it. The modern Mercury is rather dull—or should we say, a good challenge to the writer's imagination?

7.2 Venus. If Mercury was for a long time something of a mystery to astronomers, Venus was a positive embarrassment. Galileo, back in 1610, took a look at the Planet of Love with his homemade telescope and noted that the surface seemed completely featureless. That, improvements in telescopes and observing techniques notwithstanding, was the way that Venus obstinately remained for the next three and a half centuries. Venus was known to be about the same size as the Earth—a "sister planet," as people were fond of saying, coming closer to Earth than any other, and only a few hundred kilometers smaller in radius (6,050, to Earth's 6,370). But if this were our sister, we knew remarkably little about her. The length of the Venus year was determined, but not the day; and the surface was a complete and total mystery, because of the all-pervading and eternal cloud layer.

Naturally, that absence of facts did not stop people from speculating. One popular notion was of Venus as a younger and more primitive form of Earth—probably hotter, and perhaps entirely covered with oceans. The logic was simple: hotter, because nearer the Sun; and clouds meant water, so more clouds than Earth meant more water. Venus might be a steamy, swampy planet, where it rained and rained and rained.

There were competing theories. Fred Hoyle, the astronomer whom we met in Chapter 2 and will meet again in Chapter 13, speculated that Venus indeed had oceans; but according to his theory they would be oceans of hydrocarbons (the ultimate answer to a fossil fuel crisis).

Hoyle's ideas sound wild, but at least they were based on an extrapolation of known physical laws. Whereas Immanuel Velikovsky, in the early 1950s, came up with the wildest, least scientific—and most popular—theory of all. Venus, he said, was once part of *Jupiter*. By some unspecified event it was ripped out of the Jovian system and proceeded inward. There, after a complicated game of celestial billiards with Mars and the Earth, it settled down to become Venus in its present orbit. And all this took place not at the dawn of creation of the solar system, but recently, 3,500 years ago. Among other things, the multiple passages of Venus past the Earth stopped our planet in its rotation, caused a universal deluge (the Flood), parted the Red Sea, and caused numerous other annoyances.

Read Velikovsky, by all means, for wild ideas—but don't believe him. We will mention just one problem with the theory, that it violates the law of conservation of angular momentum, and leave it at that.

In the past thirty years, space probes have dramatically changed our knowledge and understanding of Venus. The present description runs as follows:

- The period for Venus to make one complete revolution about its axis is 243 Earth days. This is *longer* than the Venus year, of 225 Earth days. Also, since the planet rotates in the opposite sense from its direction around the Sun, its day—the time from noon to noon for a point of the planet—is 117 Earth days.

Would-be world-builders please note: It is difficult to visualize the relation between the time a planet takes to rotate on its axis (the *sidereal period*), the length of its day (from noon to noon), and the length of its year. However, there is a simple formula that relates the three quantities. If R is the time in Earth days for the planet to rotate on its axis, D is the length of its day, and Y the length of its year, then $1/D = 1/R \pm 1/Y$, where the plus sign is used when the planet rotates on its axis in the *opposite* sense from its travel around the Sun. For Venus, $Y = 225$ Earth days, $R = 243$ Earth days, so $D = 1/(1/243 + 1/225) = 117$.

- The pale yellow clouds of Venus are not water vapor. They are sulfuric acid, the result of combining sulfur dioxides and water. These sulfuric acid clouds stop about 45 kilometers above the surface, and below that everything is very clear, with almost no dust. The whole atmosphere is about 95% carbon dioxide. The lighting level at the surface is roughly like that of a cloudy day on Earth, though there are frequent storms in the clouds, and lots of lightning.

- The pressure at the surface is about 90 Earth atmospheres. Such a pressure may seem to offer impossible problems for the existence of life, but that's not the case. A sperm whale, diving in Earth's oceans to deeper than a kilometer, comfortably endures a pressure of more than a hundred atmospheres—and returns to the surface unharmed a few minutes later. We still don't know how the whale is able to do that.

- Venus is *hot*. In this way the modern picture of Venus is like the old one, but it is probably hotter than anyone expected. The surface temperature is somewhere between 460 and 480 degrees Celsius, and highly uniform over the whole surface. Since the axial tilt of Venus is only about 6 degrees, there are no seasons to speak of.

  Venus is hot for the same reason that a greenhouse is hot. Solar radiation gets into the atmosphere easily enough, but longer wavelength (heat) radiation from the surface is then trapped by the thick carbon dioxide atmosphere (or glass, in the case of the greenhouse) and cannot escape.

- Thanks largely to the Magellan spacecraft, we have a high-quality radar map of almost the whole surface of the planet. (Note: We still lack such a complete radar map of the surface of the Earth.) Venus is a barren place of rocky uplifts and shallow, melted-down craters. It is nothing like the old stories; no swamps, no intelligent amphibious life forms, no artifacts but a few burned-out spacecraft from the Soviet Union and the United States. But there are mountain ranges, well-mapped by orbiting imaging radars, and a great rift valley, bigger than any other in the solar system.

There is an interesting difference between the general surface structure of Earth and Venus. If we plot the average altitude of surfaces on Earth (including the seabed) we find that there are two peaks in the distribution: they represent the ocean floor and the continental platforms, separated by about five kilometers. This two-story world is a consequence of plate tectonics, where moving plates lift the land surfaces. When we make the same plot for Venus, a different picture emerges. We have a single peak, at the most common average elevation. There are uplands, a vast rift valley, and shallow basins, but they all cluster around this one average value.

Why are plate tectonics not a major force on Venus? Here we are on speculative ground. Theorists argue that the high surface temperature gives rise to a thick, light crust, which is too buoyant to be subducted (forced under) even if plates collide. Others argue that Venus is like a very young Earth, where we have yet to see the effects of plate tectonics. In perhaps a billion years Venus will see the rise of continents, and conditions may perhaps change to ones more congenial to life.

- Venus possesses no appreciable magnetic field. This is strange, since the planet is so like Earth in size and composition. However, the lack of field may be related to the planet's slow rotation, which would greatly reduce the dynamo effects of a liquid iron core.
- There remains one general question: Why is our sister planet so different from Earth in so many ways? One possibility: The Earth has a large moon; Venus has none. More and more, the presence of the Moon seems important, although I have yet to see an authoritative and persuasive discussion of the reasons.

7.3 Earth. We will say little about our own planet. Not because there is nothing to say, but because there is so much. Although this is our home, we might still argue that our understanding of Earth *as a planet* is in its early days.

Consider just a few examples. The theory of plate

tectonics, already referred to, was geological heresy fifty years ago. Alfred Wegener proposed the theory in the early part of this century, but since he was a meteorologist rather than a professional geologist, he was either ignored or laughed away. Only when the evidence of seafloor spreading became undeniable did geologists begin to accept the ideas of plate tectonics, which today underpin almost all serious geomorphological work.

A second example is the theory of primordial methane. This proposes that methane has been present in the interior since the formation of the planet, rather than being formed recently and close to the surface by the breakdown of more complex molecules through heat, pressure, and biological processes.

A third example is the Gaia theory proposed by James Lovelock and championed by Lynn Margulis. We will discuss this in Chapter 13, and note here only that it, today, is in the same state of "scientific heresy" as Wegener's theory in the 1920s.

We know remarkably little about our own Earth—and what we "know" changes with every generation.

7.4 The Moon. Other than Earth, this must be the most familiar and best-known planet or satellite in the solar system. Humans have been looking up at the Moon and studying it for all of history. Its influence on Earth, and on each of us individually, is profound. There are lunar tides running within our bodies, just as they ebb and flow in the seas of Earth. We are very familiar with our own 24-hour circadian rhythms, and how we feel at different times of day. But we are also affected by the more subtle lunar rhythm, imposing a cycle on our bodies in ways we have still to understand.

Forty years ago, our ignorance of the Moon was quite striking. For example, the Moon always presents approximately the same hemisphere to Earth (small oscillations, known as librations, allow us to see a little more than half the Moon's surface). We had no information to tell us what lay on the far side of the Moon. A good deal of wild speculation could be tolerated. It was even possible to imagine a deep depression on the back of the Moon, where there could be an atmosphere and possibly life.

That idea went away in 1959, when a Russian spacecraft, Lunik III, took and transmitted to Earth pictures of the far side of the Moon. It looked, disappointingly, rather like the side that we already knew.

However, there were still plenty of things to speculate about. For example, the craters: were they caused by volcanoes, or were they meteor impacts? Forty years ago no one had any proof one way or the other. The flat, dark "seas" on the Moon: they were certainly not water, but might they be deep dust pools, ready to swallow up any spacecraft unwise enough to attempt to land on one of them?

Today we have many of the answers. First, we know that the surface of the Moon is old. The measured ages of lunar rock samples brought back in the Apollo program are in the billions of years. Half of them are older than any rocks ever found on Earth. Even the "new" craters, like Tycho, measure their ages in hundreds of millions of years. The dust pools are not there. Astronauts who landed on the Moon reported a layer of dust, but no sign of the deep, dangerous seas of an earlier generation's speculations.

The Moon is of great interest to scientists; but it seems fair to say that to most people it is a dull place. There are no known substantial deposits of valuable minerals, no air, little water. The Clementine spacecraft, according to a widely reported Defense Department press release, in 1993 "discovered" a lake of ice in a crater near the north pole of the Moon. However, the actual scientific paper in *Science* concerning the radar signals was far more circumspect, and merely noted that Clementine's radar return signal was *consistent* with the presence of water. A 1998 observation by the Lunar Prospector spacecraft made newspaper headlines with the announcement that a hundred billion tons of water had been found on the Moon. The most impressive thing to me is how *little* water that is. It is a small pond, ten feet deep and seven miles across. On Earth it would hardly be noticed.

Human colonies on the Moon seem possible within a generation, but they may exist mainly to send materials back out into space, or to take advantage of the radio quiet zone on the lunar far side (we flood the near side, and most of space, with our incessant babble). The biggest advantage of the Moon may turn out to be its low escape velocity, only 2.4 kms/sec, allowing cheap shipment of materials from the Moon to Earth orbit.

I do not think that a lunar base will satisfy our urge to develop the planets. The Moon is too much an offshore island of Earth. We have already paddled our dugout canoes there a few times, and we will be going back. But it is not our new continent, our "new-found-land."

That new-found-land may be Mars.

7.5 Mars. The Red Planet has had some bad publicity over the years, in science fictional promises that were not kept.

There were the canals of Mars, which Percival Lowell thought he could see very well and believed were of artificial origin, but which other people had trouble seeing at all.

And of course there were the Martians, given very poor press by H.G. Wells in *The War of the Worlds* (Wells, 1898). They were sitting up there on Mars, with their "vast, cool, and unsympathetic" minds set on taking over Earth.

Regardless of whether the Martians were good or bad, at the turn of the century almost everyone agreed that there was life on Mars. Although Venus is Earth's sister planet, from many points of view Mars is a more convincing Earth look-alike. It has a day just a few minutes longer than a day on Earth (24 hrs., 37 mins.). It has an axial tilt almost the same as Earth's, so the cycle of the seasons should be similar. And it has an observable atmosphere, although one that a generation ago was of unknown composition and density. There are noticeable seasonal changes in both the planet's color and the size of the polar caps with each Martian summer and winter.

Intelligence, maybe; life, a sure thing. That seemed to be the common attitude toward Mars eighty years ago.

And the modern Mars? No canals, but a cratered sand-worn surface that looks more like the Moon than Earth. Months-long sand storms. No surface water, but lots of signs of ancient water run-off. Stupendous mountains, twice as high as any on Earth; a vast canyon (*Vallis Marineris*) that would easily swallow the Grand Canyon whole; and plenty of jagged surface rocks. That was the report that came back from the Mariner, Mars (Soviet) and Viking spacecraft, and also from the Viking Lander. In 1976 the Lander also looked for life with its onboard experiment package. The first results were outstandingly positive, too good to be true—there seemed to be chemical indicators of life everywhere. Then the investigators decided, yes, those results are too good to be true, and they're not true.

The most widely held view, prior to August 1996, was that Mars lacked life completely and probably never had it. That situation changed dramatically with the NASA announcement that analysis of a meteorite found in Antarctica revealed possible evidence of ancient single-celled life on Mars. The 1997 Pathfinder lander, and its roving companion Sojourner, were not designed to look for life, though they did find more evidence of long-ago surface water.

The current Mars atmosphere is not promising to support the forms of life that we know best. The pressure at the surface is only one percent of an Earth atmosphere, and it is mostly carbon dioxide and nitrogen. Surface temperatures range from the freezing point of water, at low points on the equator at high noon, to $-100°C$ or colder. That is not most people's idea of a mild climate. On the other hand, there are terrestrial organisms that can stand those temperatures, and even thrive if they have access to water. And there *is* water on Mars. It is found in the polar caps, believed to be a mixture of water ice and solid carbon dioxide ("dry ice"). Some analyses also find evidence for deep liquid water, an idea developed in detail in Kim Stanley Robinson's monumental trilogy, *Red Mars, Green Mars, Blue Mars* (Robinson, 1993, 1994, 1996). Before you consider writing about Mars colonization, read Robinson's work.

In spite of everything, humans could live on Mars. The available land area is roughly equal to the land area of Earth. The atmosphere is dense enough to be useful for aerobraking spacecraft, or flying an aircraft. The low gravity, only $^2/_5$ of Earth gravity, helps a lot. If there are no Martians now, someday there will be.

7.6 The moons of Mars. Mars has its own moons, two of them. However, if attention to objects in the solar system were to be given in proportion to their size, Phobos and Deimos would be totally ignored. They are tiny objects, each only tens of kilometers across.

In Chapter 1 we mentioned Jonathan Swift's 1726 "predictions" of the existence and major characteristics of these moons, long before there was any chance of discovering them. The little moons themselves would not be discovered for another century and a half. They were finally seen by Asaph Hall, in 1877. Later observations, between 1877 and 1882, gave estimates of their distances from Mars and their orbital periods.

Until forty years ago, distances from Mars and orbital periods were all that anyone knew of Phobos and Deimos. In 1956, Gerald Kuiper estimated their diameters, giving figures of 12 kms for Phobos and 6 kms for Deimos. But the real quantum leap in our knowledge had to wait until 1977, one hundred years exactly after Asaph Hall's discovery. In that year, the Viking 2 spacecraft took a close-up look at both moons.

Neither Phobos nor Deimos is anything like a sphere. They are ellipsoids of roughly similar shape. Phobos is 27 by 21 by 19 kilometers, and Deimos 15 by 12 by 11 kms. They are both tidally locked to Mars, so that they always have their longest axes pointed towards the planet. They have battered, cratered surfaces, and Phobos has one huge crater, Stickney (named after Asaph Hall's wife, Angeline Stickney, who encouraged him to keep looking for the moons when he was ready to give up). Stickney is about ten kilometers across—nearly half the size of the moon-let. Both moons have a regolith, a dusty surface layer of fine-grained material, and both are thought to be captured asteroids. There is some suggestion that Phobos may have water locked within it, because some of its surface features suggest steam has escaped there after past meteor impacts. Phobos looks more and more like a tempting target for anyone interested in conducting a manned Mars expedition, perhaps in the first decades of the twenty-first century. With its low gravity and location, it is an equally good target for science fiction writers.

7.7 The asteroid belt. This is also good frontier territory for speculation. "Asteroid" means "having the form of a star" and it is a terrible name for what are, in essence, small planets. "Planetoid" would be much better. Unfortunately, we seem to be stuck with the word, and also with "asteroid belt." There is a huge number of asteroids, ranging from the biggest, Ceres, at 974 kilometers diameter, through Pallas (538 kms diameter), Vesta (526 kms), Juno (268 kms), and on down to boulders and pebbles. We still know little about most of them, beyond their shapes, rotation periods, and light-reflectance curves. We have had close-up photos of two (Gaspra and Ida, the latter a double asteroid of two bodies, Ida and Dactyl, bound to each other by gravity), and we have Hubble Telescope images of Vesta and other large asteroids.

Some asteroids have left the main belt, between Mars and Jupiter, and swing in on orbits much closer to the Sun. This class of so-called *Earth-crossing asteroids* includes its own subgroups: the *Apollo* asteroids have orbits crossing Earth's orbit; the *Aten* asteroids are on average closer to the Sun than is the Earth (their semimajor axis is less than Earth's); and the *Amor* asteroids cross the orbits of both Earth and Mars. Finding such asteroids is today an active business, because it takes less fuel to get to them from Earth than to most other places in the solar system. Many contain valuable minerals. A small, metal-rich asteroid, maybe a mile across, should provide as much nickel as all Earth's known commercial deposits, and in quite a pure form. Don Kingsbury's "To Bring in the Steel" (1978) tackled the theme of mining one.

People have proposed other uses for Earth-crossing

asteroids. Moved to Earth orbit (feasible if the necessary volatile material for fuel can be found on the asteroid itself), such bodies could be used to protect other satellites and installations, or as a threat to ground-based facilities.

There is an old controversy surrounding the asteroids: Are they fragments of matter that never got together to form a planet, or were they once a planet that for some reason catastrophically disintegrated? Forty years ago, no one could offer firm evidence one way or the other. Today, most astronomers argue that the planet never formed. Jupiter's powerful gravitational field prevented the separate bodies from ever coalescing.

However, there have been other opinions. In 1972, the Canadian astronomer Ovenden examined the rate of change of planetary orbits, and concluded that they are varying too rapidly for a solar system that has supposedly been fixed in major components for hundreds of millions of years. Ovenden looked at the changes, and found they were consistent with the disappearance from the system of an object of planetary dimensions in the fairly recent past. He concluded that a body of about 90 Earth masses (the size of Saturn) had vanished from the solar system about sixteen million years ago. Three years later, Van Flandern at the U.S. Naval Observatory analyzed the orbit of long-period comets. He found many with periods of about sixteen million years, and they seemed to have left the solar system from a particular region between the orbits of Mars and Jupiter.

Where do I stand on this question? Reluctantly, I conclude that the asteroids were never a single body. They date back to the origin of the solar system, and have probably existed in their present form ever since.

On the other hand, in my novel *Sight of Proteus* (Sheffield, 1978), a planet between Mars and Jupiter blew itself apart and created the asteroid belt. If I could get away with it, why shouldn't you? You can do as I did, and cite Ovenden and Van Flandern.

7.8 Jupiter. It is convenient to break the discussion of the planets of the solar system into two parts: anything closer to the Sun than the asteroid belt, and anything farther out. This division is also logical. The inner system contains small, dense, rocky bodies, of which Earth is the biggest and heaviest. The outer planets are (except for Pluto, which is probably not a true planet at all) large and diffuse gaseous bodies, with little or no solid core.

Until the invention of the telescope, what we knew about the outer solar system could be summarized very simply: it was Jupiter and Saturn, seen only as specks of light in the sky.

This, even though Jupiter is by far the biggest planet of the solar system, a bully whose gravitational field grossly perturbs every other body orbiting the Sun. With a diameter eleven times Earth, and a mass 320 times as big, Jupiter contains more material than all the rest of the planets put together. Its density was estimated more than a century ago, at 1.3 grams/cc. This is a low value compared to Earth, so astronomers knew that Jupiter must contain a large fraction of light elements.

Jupiter was known to be in rapid rotation, spinning on its axis once every ten hours. This, together with its great size, means that it bulges noticeably at the equator. The equatorial radius is about 6 percent bigger than the polar radius.

The Great Red Spot on Jupiter was observed in the seventeenth century (first noted by Robert Hooke, in 1664). The feature has dimmed and brightened over the years, but it is known to have been there continuously since at least 1831. It has been observed regularly since 1878. The size varies quite a bit. At the beginning of this century it was about 45,000 kms by 25,000 kms, twice today's size. But even in its present shrunken state, the Great Red Spot could easily swallow up Earth.

Forty years ago the nature of the Great Red Spot was quite unknown. One theory, still acceptable in the 1940s, held that the Spot was a new satellite of Jupiter in the process of formation, ready to split away from its parent planet (shades of Velikovsky). Other later ideas, from the 1960s, include a floating island of a particular form of water-ice (a phase known as Ice VII), or an atmospheric cloud cap over a deeper floating island. The spot moves around on the surface of Jupiter, so it certainly has to be a floating *something*.

The other long-observed features of Jupiter were the striped bands that circle the planet parallel to the lines of latitude. Their appearance also suggested clouds. Given Jupiter's low density, those clouds were assumed to be very deep, but their composition was largely a matter of guesswork and something of a mystery. Speculation based on the composition of the Sun suggested that Jupiter ought to be mainly hydrogen and helium, but the direct observations of the 1960s showed only methane and ammonia.

It has been known since the 1950s that Jupiter is an intense emitter of radio noise, but the mechanism for its production was vague. It was known that somehow it seemed to correlate with the position of Io.

As for satellites, in 1960 a round dozen of them were known. These included the four major ones discovered by Galileo in that marvelous year of 1610 when he first applied his telescope to astronomy. Now termed the Galilean satellites, they are, in increasing distance from the planet, Io, Europa, Ganymede, and Callisto. In 1892 a fifth satellite was found, inside the orbit of Io. It was named by its discoverer, E.E. Barnard, simply "V," the Roman number for five. Later it became known as Amalthea. The other satellites, all more distant than Callisto, were numbered in the order of their discovery. Other than size estimates and orbit parameters, not much was known about any of the moons of Jupiter in 1960. The larger ones showed a few light and dark spots, and none seemed to have an atmosphere. The four outermost moons are much farther from Jupiter. They are in retrograde orbits, i.e. they are moving around Jupiter in the opposite direction from the planet's spin, and they were generally thought to be captured asteroids.

Today's picture of the Jovian system, thanks largely to observations by the Pioneer, Voyager, and Galileo spacecraft, is vastly different from that of even thirty-five years ago. The satellites that were then little more than points of light are now well-mapped worlds, each moon with its own unique features and composition. The atmosphere of Jupiter itself has been looked at in great detail, and it is known to contain complex churning cloud patterns, with infinitely detailed vortices. The Great Red Spot has given up its secrets: it is a vast semipermanent storm system, a hurricane fueled by Jupiter's rapid rotation and lasting for hundreds of years.

We still know less than we would like to about Jupiter's interior. The escape velocity from the planet is about 60 kms/sec, and once you go there it is hard to get away. The present picture of the planet's interior is of a deep, slushy ocean of hydrogen under fabulous pressure. At three million Earth atmospheres, seventeen thousand kilometers deep in Jupiter's atmosphere, hydrogen is believed to change to a metallic form. Deep below that is perhaps a small central core of rock and iron about the size of the Earth.

We now have confirmation that Jupiter is composed largely of hydrogen and helium, with an observed 19 percent helium in the upper atmosphere. And we have confirmation that Jupiter gives off more energy than it receives, a result that was still tentative twenty-five years ago. Since the planet is a net emitter of energy, that energy must be produced somewhere in the deep interior. And there must be adequate convection mechanisms to bring the heat to the outer layers. In fact, Jupiter is almost a star; a bit bigger, and it could support its own fusion reactions.

Jupiter has electric and magnetic fields in keeping with its size. The powerful magnetic field captures and accelerates the "solar wind," the stream of energetic charged particles emitted by the Sun. As the nearest large moon, Io, moves through that swarm of particles it generates and sustains a "flux tube," a tube of current, five million amperes strong, that connects Io and the atmosphere of Jupiter. This in turn stimulates intense electrical activity in the Jovian cloud systems. The cloud tops seethe with super-bolts of lightning, and they generate powerful radio emissions from the planet. The night side shimmers with auroras, also observed by the electronic eyes of the Voyager spacecraft in their 1979 inspection of the planet.

The Voyager and Galileo spacecraft sent back quite extraordinary images of the major moons of Jupiter. Amalthea, the smallest and nearest-in of the previously known Jupiter satellites, proved to be a lumpy, irregular ellipsoid, about 265×170×155 kms. The longest axis always points towards Jupiter. Amalthea is tidally locked to face the parent planet.

Io, the next one out, is tidally locked also. Io is a spectacular sight. It looks like a smoking hot pizza, all oranges and reds and yellows. As it sweeps its way through that high-energy particle field surrounding Jupiter, tidal forces from the parent planet and its companion satellites generate powerful seismic forces within it. Io is a moon of volcanoes. Many active ones have been observed, spewing out sulfur from the deep interior.

Europa is my own favorite of the Galilean satellites, and much of my novel *Cold as Ice* is set there. Europa is the smallest of the four, with a mass about ²/₃ of our own Moon. And it seems to be an ice world. There is a smooth, flat surface of water-ice, fractured by long linear cracks, ridges and fissures. Underneath those there is probably liquid water, kept from freezing by the tidal heating forces from Jupiter and the other Galilean satellites. Europa has an estimated radius of 1,565 kms, and an estimated density of 3 grams/cc. It is believed to possess a rocky silicon core, with an outer ice/water layer maybe 100 kilometers thick. There has been speculation, some of it mine, that the ice-locked waters of Europa could support anaerobic life-forms. These would derive their energy from hydrothermal ocean-floor vents, much like similar life-forms in Earth's deep oceans.

Ganymede is the biggest moon in the solar system, with an estimated radius of 2,650 kms. It has a low density, about 1.9 grams/cc, and is thought to be about 50 percent water. The brightness of Ganymede's surface suggests that it may be largely water-ice. The surface is a mixture of plains, craters, and mountains, not unlike the Moon.

Callisto, the outermost of the Galilean satellites, is all craters—the most heavily cratered body in the Jovian system. It has a radius of about 2,200 kms, slightly smaller than Ganymede and Saturn's biggest moon, Titan. It has the lowest density of any of Jupiter's moons, again suggesting that we will find lots of water-ice there. The surface of Callisto seems very stable. It has probably not changed much in four billion years, in contrast to Io's fuming surface, which changes daily.

As for the other satellites of Jupiter, we still know little about them. However, the Voyager mission did add one to their number—a small one, less than 40 kms across. That moonlet orbits at the outer edge of Jupiter's ring system.

All this, and rings too? Yes. Twenty years ago, Saturn was thought to be the only ringed planet. Now we know that Jupiter, Uranus, and Neptune are all ringed worlds. Jupiter has a thin ring, well inside the orbit of Amalthea. It has a sharply defined outer edge, and it sits about 120,000 kms out from the center of Jupiter.

TABLE 7.1 (p. 185) shows a "score card" of the moons of Jupiter.

7.9 Saturn. Saturn is about twice as far as Jupiter from the Sun (and hence from us—as most solar system distances go, we sit very close to the Sun). Saturn is a little smaller (58,000 kms radius, to Jupiter's 70,000); and since it is farther from the Sun it is less strongly illuminated. For all these reasons, Saturn is more difficult to observe from ground-based telescopes, and our knowledge of a generation ago reflected that fact. The most famous feature of Saturn is the ring system. Those rings were first observed, like so much else in the solar system, by Galileo in 1610, but he was baffled by them and had no idea what they might be. Huygens, working forty-five years later with a better telescope, was the first person to deduce the nature of the rings. Nearly two hundred years after that, in 1857, Maxwell showed on mathematical grounds that the rings could not be solid. They have to be a swarm of some kind of particles. However, the size and composition of those particles were unknown even as recently as twenty-five years ago, although the popular theory was that they were small chunks of ice. The rings of Saturn were imagined as snowballs, of varying sizes.

It was known that there was not one ring, but several. In 1675 Cassini observed at least two rings, separated by what we now call the Cassini division. A third ring, the Crape ring, was observed in 1838, and again in 1850.

As for the planet itself, Saturn seemed a smaller, lighter version of Jupiter. Its radius was close to Jupiter's, but its density was only 0.7 grams/cc (it is the least-dense large body in the solar system; Saturn would float in water, if you could find a big enough bathtub. Presumably it would leave a ring).

Saturn weighs in at 95 Earth masses, versus 320 for Jupiter. The surface shows the same banding as Jupiter's, but with less visible detail. The equatorial bulge is even more pronounced, with a polar radius of 54,000 kms and an equatorial radius of 60,000 kms. The planet's volume is about 750 times that of Earth, and the rotation period is 10 hours and 15 minutes (although that period is not the same at all latitudes; Saturn rotates faster at the equator than near the poles). Saturn's axis is inclined at 26.75 degrees to its orbit, so that unlike Jupiter it has substantial "seasons."

By 1960, nine satellites of Saturn had been discovered. In order, moving outward from the planet, these are Mimas, Enceladus, Tethys, Dione, Rhea, Titan, Hyperion, Iapetus, and Phoebe. Percival Lowell thought he had seen a tenth one in 1905, and he named it Themis, but he had no more luck here than he did with the canals of Mars. No one else has ever seen it.

Today, thanks again mainly to the Voyager spacecraft, we know that the atmosphere of Saturn is mostly hydrogen, with rather less helium than Jupiter (11 percent above the clouds, versus 19 percent for the larger planet). Methane, ammonia, ethane, and acetylene have also been observed in the atmosphere; and like Jupiter, Saturn gives off more energy than it receives from the Sun, so there must be internal sources of heat. The clouds of Saturn show a number of long-lived features, including atmospheric cyclonic patterns like the Great Red Spot on

Jupiter. Saturn at the time of the 1981 Voyager 2 encounter had nothing of that size, though it did have one red spot about 6,000 kms long in its southern hemisphere. However, in September, 1990, a new "Great White Spot" was found on Saturn by the ground-based observations of amateur astronomers. Images taken by the Hubble Space Telescope revealed that this feature is a huge cloud system, extending a third of the way around Saturn's equator. Its cause and its degree of permanence are unknown.

The rings of Saturn are known to be infinitely more complex than anyone dreamed of twenty-five years ago. There are not two or three rings but thousands of them, each one very narrow. And they are not just simple rings. Sometimes there are radial gaps in them, "spokes" that come and go within a period of a few hours. Some of the rings are interwoven, plaited together in ways that seem to defy the laws of classical celestial mechanics. (They don't, but they do call for nontraditional techniques of orbital analysis.) Other rings are "herded" along in their orbits by small shepherding satellites that serve to control the location of ring boundaries. The composition of the rings has been confirmed. They are indeed mostly water-ice—bands of snowballs, hundreds of thousands of kilometers across.

The count of satellites for Saturn, not including the rings which are themselves composed of innumerable small satellites, has gone up substantially. Eighteen have been named. Not surprisingly, the new satellites do tend to be on the small side, although one of them, Janus, circling Saturn at about 150,000 kms distance, is comparable in size with Phoebe.

Of all these moons, Titan has received the most attention. We know that it has a substantial atmosphere, with a surface pressure of 1.6 Earth atmospheres. It is composed mainly of nitrogen, with a good fraction of methane (as much as 10 percent down at the surface, and less higher up). The dark-red color of Titan is due to a photochemical smog of organic (i.e., carbon-containing) compounds, and ethane, acetylene, hydrogen cyanide, and ethylene have all been detected. The surface temperature has been measured as about −180°C. One plausible current conjecture is that Titan has an ocean—but an ocean of ethane and methane, rather like liquefied natural gas. All water on Titan will be well-frozen, but water-ice may lie below that frigid sea. Just as the old canals of Mars seem to have appeared as linear features on Europa, the petroleum oceans of Venus may be here, on Titan.

The rest of the satellites are much smaller, devoid of all signs of atmosphere, and their low densities suggest that they contain a good deal of water-ice. All the known moons are cratered, and Mimas has one gigantic crater on it, nearly 130 kms across. Iapetus shows dark-red material on its leading face, suggesting that water-ice may have been eroded from that hemisphere by meteor impact as the moon moves in its orbit around Saturn. Another possible explanation is that water-ice has been preferentially deposited on the trailing hemisphere.

The "score card" for Saturnian satellites is given in TABLE 7.2 (p. 186). The surface radius of Titan, 2,575 kms, makes it a little bit smaller than Ganymede. It is still bigger than Callisto or any other moon in the solar system.

7.10 Uranus. Until 1781, the solar system ended at Saturn. William Herschel's discovery of Uranus changed that forever; now no one is sure where the "edge" of the solar system should be placed.

Uranus, smaller than Saturn and almost twice as far from the Sun, revealed few of its secrets to ground-based telescopes. The "day" on Uranus was poorly determined even thirty years ago, estimated as anything from 10.5 to 18 hours. The large uncertainty in that number stemmed from an inability to see *any* features on the Uranus surface by ground telescope observation.

Soon after the planet was discovered, it was learned (by observing the moons of Uranus) that the rotation axis is highly tilted relative to the orbital plane. The planet progresses around the Sun "on its side" like a rolling ball. Other than the size (about 25,000 kilometers estimated radius) and color (greenish, suggesting an atmosphere of hydrogen and helium plus methane and

ammonia) not much more was known about the planet.
The images of Uranus obtained by Voyager 2 in 1986
were something of a disappointment. The planet resem-
bled a hazy billiard ball, with scattered high-lying clouds,
probably of methane. The rotation of those clouds, plus
direct observation of a rotating magnetic field (a source
of observations previously quite unavailable) yields a
Uranus day of 15.6 hours.

That rotating magnetic field is one of the most
interesting facts about the planet. It is sizable (0.25 gauss
at the planet's surface, compared with 0.31 gauss for
Earth) and it is markedly off-axis compared to the planet's
rotation. For Earth, Jupiter, and Saturn, the magnetic field
axis and the rotation axis point in almost the same
direction. For Uranus, they are inclined at 55 degrees
to each other.

Analysis of atmospheric composition shows Uranus to
be between 10 and 15 percent helium, much the same
as Jupiter. Heat balance calculations confirm that Uranus
lacks any internal source of heat.

Let us move from the planet itself, to the objects that
orbit around it.

Before 1977, Saturn was believed to be the only ringed
planet. In that year rings around Uranus were discovered
by ground-based observation (stars disappeared and
reappeared when the rings of Uranus were passing in
front of them). Voyager 2 showed that all the rings are
narrow and extremely dark in color; thus they cannot
be water-ice like Saturn's rings. The pattern of scattered
light from the rings suggests that there is little fine dust
in them, which makes them quite unlike the rings of
Saturn. That might be due to the off-axis magnetic field.
Small particles with a high charge-to-mass ratio could
be cleared out of the rings by the regular magnetic field
variation, so only the larger particles would be left. Six
of the rings appear elliptical, which was unexpected and
suggests that they may have been created recently
(speaking in astronomical terms; i.e. no more than a few
million years ago).

The search for moons around Uranus began as soon
as the planet was discovered. The biggest two, Titania

and Oberon, were discovered by Herschel himself in 1787. And from 1851–52, William Lassell found two more, Ariel and Umbriel. No one else saw those two for over twenty years, and many must have wondered if they really existed; but Lassell was at last proved right. The fifth and final one of the "old" set of moons (those known before the Voyager flyby) was discovered in 1948 by Gerald Kuiper. It was named Miranda.

Today, 15 moons of Uranus are known and named. The new ones are between 13 and 77 kilometers in radius. We know little of their surface detail or composition. However, high-resolution images are available of Miranda, Ariel, Umbriel, Titania, and Oberon.

The score card for the moons of Uranus is given in TABLE 7.3 (p. 187). Note that all the newly discovered small moons are closer to Uranus than the five previously known. The bigger moons show more evidence of internal activity than anyone expected, though at −210°C they are even colder than the pre-Voyager estimate of −190°C. They reveal what appear to be old impact craters, fault structures, and newer extruded material in crater floors. The exception is Umbriel, which displays a bland, dark, featureless disk.

Voyager 2 came within 29,000 kms of Miranda's surface, the spacecraft's closest approach to anything in the Uranus system. The images of that moon show an object with unexpectedly complex and inexplicable surface geology. For a first-rate science fiction story set on Miranda, try G. David Nordley's "Into the Miranda Rift" (Nordley, 1993).

7.11 Neptune. Unlike the other planets of the Solar System, which first appeared to humans as bright points of light in the night sky, Neptune was not discovered by observation. It appeared as an abstract deduction of the human mind.

The planet showed its presence in the first half of the nineteenth century as a small anomaly, a difference between the calculated and observed position of Uranus in its orbit. An Englishman, John Couch Adams, and a Frenchman, Urbain Le Verrier, took that small discrepancy, solved (independently) a difficult celestial mechanics problem of "inverse perturbations," and correctly predicted the existence and location of Neptune. When the planet was observed in 1846, to many people of the time it must have seemed like a magic trick. A paper-and-pencil calculation, unrelated to the real world, had somehow told of the existence of a new planet. This was mysterious, even mystical. When Gustav Holst composed his orchestral suite, *The Planets*, he labeled Neptune as "The Mystic" and wrote music to match.

Neptune has a mean distance from the Sun of 4.5 billion kilometers and a period (the Neptunian year) of almost 165 years. The great distance makes Earth-based observations extremely difficult. Light takes four hours to travel from Neptune to Earth. Out at Neptune, the Sun subtends only one minute of arc in the sky, and the intensity of sunlight is one nine-hundredth of what we experience here.

The Voyager 2 encounter revealed Neptune's equatorial radius to be 24,700 kms (since Neptune does not have a solid surface, this is taken as the radius where the pressure equals one Earth atmosphere). Neptune has a mass 17 times that of the Earth, and an average density of 1.64 grams/cc. The Neptunian day was revised to 16.11 hours, based on the rotation of the planet's magnetic field. That magnetic field is substantial, and its axis is offset 47 degrees from the planet's axis of rotation. In addition, the center of the magnetic field does not coincide with the planet's center of mass. As a result the field at the surface ranges from less than 0.1 gauss in the northern hemisphere to more than 1 gauss in the southern.

The appearance of the planet itself is striking. Unlike bland Uranus, Neptune shows atmospheric detail more like Jupiter and Saturn. There is a Great Dark Spot of midnight blue, calling to mind the Great Red Spot of Jupiter, and around the spot are bright, cirrus-like clouds that move along lines of latitude. This atmospheric activity may be a consequence of a net heat outflow, for like Saturn and Jupiter but unlike Uranus, Neptune gives off more energy than it receives from the Sun; in this case about 2.7 times as much. The minimum observed temperature on Neptune is a frigid 50 Kelvin, up near the top of the atmosphere.

Earth-based observations of Neptune, plus theoretical arguments, had suggested that its atmosphere would be hydrogen and helium with some methane. That has been confirmed. The helium is about 15 percent of the total, and small amounts of both methane and acetylene were found.

In the mid-1980s evidence had been found of rings around Neptune based on ground observations; or rather, there seemed to be evidence of *partial* rings. The way to find rings is to look for a star dimming and then brightening again, just before the planet passes in front of it. If there is a ring, then the same thing should happen again when the star reappears on the other side of the planet. This *stellar occultation* method was used for Neptune, just as was done in the case of Uranus.

However, although applying the technique to Neptune sometimes gave a dimming of the star for a couple of seconds, and a brightening before it vanished from sight behind the planet, there was no dimming when it reappeared!

In any event, full rings were found during the Voyager 2 encounter. There are three complete rings, and an outermost ring containing three bright, dusty arcs within it. These ring arcs caused the peculiar occultation results found in the earlier ground-based measurements.

Before Voyager 2, Neptune had two known satellites. The larger, Triton, was found in 1846 by that remarkable observer and discoverer of Uranus's Ariel and Umbriel, William Lassell, just ten days after the discovery of Neptune itself. Triton is big, with a radius of 1,350 kms, and has about a third of our Moon's mass. It travels in a retrograde orbit, opposite to the direction of planetary rotation. It has a period of 5.9 days, inclined at 23 degrees to the Neptune equator.

Nereid, the second satellite, is much smaller. It was discovered by Gerald Kuiper in 1949, and it travels in a very elliptical orbit, far out from the planet, with a period of 360 days. It and Triton are almost certainly captured bodies, caught in Neptune's gravitational net.

The Voyager encounter added half a dozen to the count of Neptune's moons. I have a personal fondness for Proteus, the biggest of these moons. Proteus is shaped like a knobby apple, and it may be the largest highly asymmetrical body in the solar system. Not much is known about it. Proteus orbits close to Neptune, where its own reflected light is overpowered by the light of its primary.

As for Triton, it is bright, and it is *cold*. The surface temperature of 38 Kelvin is the lowest measured for any body in the solar system. Nitrogen is solid at this temperature, and so is methane. The atmosphere is very thin, surface pressure between 10 and 20 millionths of an Earth atmosphere, and it is mainly nitrogen vapor with a little methane.

Any disappointment at Triton's cold, thin atmosphere is more than made up for by the satellite's astonishing surface. It possesses active geysers, "cryovolcanoes" that blow icy plumes of particles tens of kilometers high. The surface is fantastically cracked and complex, much of it showing meteorite impact craters crisscrossed by ridges of viscous material in a pattern that the Voyager team termed "cantaloupe terrain."

The score card for Neptune's moons is given in TABLE 7.4 (p. 189).

7.12 Pluto and the limits of the solar system. This planet has never been visited by any probe, so it is still wide open for science fictional conjecture. Discovered by Clyde Tombaugh in 1930, Pluto is described in most astronomy textbooks as "the most distant planet from the Sun." Actually, from 1979 to 1999, Neptune was the most distant known planet. For part of its eccentric orbit, Pluto moves within the orbit of Neptune.

Pluto's best images have been gained by the Hubble telescope. The planet has a mean radius of 1,140 kilometers. Its average surface temperature is about 43 Kelvin. There is some evidence that the surface is partly covered with methane ice, and it is conjectured that, like Triton, which it resembles in size and distance from the Sun, Pluto may have a coat of solid nitrogen.

Pluto, smaller than some satellites of Jupiter and Saturn, surprisingly has a moon of its own. Discovered in 1978 from ground-based observations, it is named Charon. It is about 590 kms in radius. Since Pluto itself is only 1,140 kms in radius, relative to the size of its planet Charon is the largest moon in the solar system. Pluto and Charon orbit each other in 6.4 days, and are 19,400 kilometers apart. The discovery of Charon allowed a good estimate of the mass of Pluto itself. That mass turns out to be small indeed, about one five-hundredth of Earth's mass. Charon's mass is still less, only one-seventh that of Pluto.

Might there be a "tenth planet," out beyond Neptune and Pluto? The search for such an object has been proposed, because one reason for seeking Pluto was a slight discrepancy between Neptune's observed and computed positions. However, after Pluto was discovered its faintness indicated that it could not be massive enough to cause the observed differences. Hence the search for "Planet X."

No such single planet has been found, but more than thirty small bodies—planetoids, minor planets, large comets, or whatever we choose to call them—have recently been discovered beyond the orbit of Neptune. They range in size from a hundred to four hundred kilometers in diameter, and are believed to be members of the *Edgeworth-Kuiper Belt*. This is often called the *Kuiper Belt*, but its existence was first suggested by the Irish astronomer K.E. Edgeworth in 1943. The EK Belt is believed to extend to at least twice the distance of Neptune from the Sun, but detection of its more remote members is extremely difficult because of the distance and low illumination levels there. The EK Belt is believed to be the source of many of the short-period comets that from time to time visit the inner solar system.

Even with the Edgeworth-Kuiper Belt, we are not at the "edge" of the solar system. In 1950, the Dutch astronomer Jan Oort suggested a source for the long-period comets. Oort proposed that there must be a vast "cometary reservoir," somewhere far out in space.

The roughly spherical Oort Cloud of comets drifts around

the Sun, weakly bound by solar gravitational attraction. Sometimes a comet will be perturbed by another star, or perhaps by a close encounter with another Cloud member. Then its orbit will change, and it may fall in toward the Sun and become visible to us. Clearly, if comets are fairly common occurrences, there must be a lot of them in the cloud. Estimates put the number in the Oort Cloud as somewhere between a hundred billion and a trillion. Each comet is thought to be a loose aggregate of water, gravel, and other volatile substances such as ammonia and hydrocarbons—the "dirty snowball" theory introduced by Fred Whipple in 1950. The Oort Cloud is a great setting for stories. I put my novel *Proteus Unbound* out there, and had lots of fun with it.

The Oort Cloud is believed to extend as far as fifteen trillion kilometers from the Sun. Fifteen trillion kilometers takes us more than a third of the way to the nearest star. Are we finally at the "edge" of the solar system?

Well, there is still Nemesis. This highly hypothetical "dark companion" to the Sun is supposed to return every 26 million years, to disturb the solar system and shower us with species-extinguishing comets that fall in from the Oort Cloud.

The existence of Nemesis is highly controversial, and I find the arguments for it unpersuasive. However, few explanations are available for periodic large-scale species extinctions. At its most distant point from the Sun, Nemesis would be almost three light-years away. At that distance, it would hardly be gravitationally bound to our Sun at all. Should it be discovered (it may be very faint, because if its mass is small enough it will not sustain its own fusion reactions), then the size of the Solar System has expanded in two hundred years from the orbit of Saturn, one and a half billion kilometers from the Sun, to the thirty trillion kilometers limit of Nemesis's orbit.

If all the natural bodies of the solar system are not enough as possible homes, there remains the possibility of making more in open space. One approach to the construction of such space colonies is discussed in Chapter 8.

7.13 Planets around other stars. Although humans can live in space and will do so in increasing numbers, planets are likely to remain our preferred home. In our own solar system, Mars is the most tempting new prospect. If Europa's water ocean exists, then that moon of Jupiter will be an equally attractive goal.

But what about more distant planets, around other stars? Do they exist? And if so, are they likely to be suitable for the development of life?

Science fiction writers have always assumed that the answer to all these questions was a definite and unambiguous *Yes!* In half the stories you will ever read, or movies and TV shows you will ever watch, it is assumed that planets exist around other stars, that they are suitable for life, and that they nurture intelligent life. Many of the intelligent life-forms are human-like to the point of ludicrous implausibility. Yet, up to 1996, there was no firm evidence at all that even one planet existed around any star other than Sol.

Certain properties of any such planets could be inferred, even if none had been observed. For example, no matter what shape a planet starts out at the time of its formation, gravitational forces will tend to make it spherical over time. When a planet happens to be rotating fast, like Jupiter or Saturn, centrifugal forces will give it a bulge at the equator. This oblateness, as it is called, is greater for Saturn than for any other planet in the solar system, but our eyes still see the disk of Saturn as circular. Anything big enough to be called a planet must be roughly spherical in shape.

For a spherical planet, the escape velocity at the surface (the speed of an object needed to escape from the planet completely) depends on only two things: the mass and the radius. Although internal composition—the way matter is distributed inside—will have a small effect, the escape velocity, V, will be close to $\sqrt{2GM/r}$, where M is the mass in kilograms, r the radius in meters, and G is the universal gravitational constant, equal anywhere in the universe to 6.672×10^{-11}. Here V is given in meters/sec. For example, in the case of

Earth, $M=5.979\times10^{24}$, $r=6,378,000$ and we find $V=11,180$; i.e., 11.18 kms/sec.

Escape velocity is important, and not only because it tells us what speed a rocket needs to get clear of Earth's gravity. It is also one of two key variables that decide whether or not a planet can hold on to an atmosphere. The other variable is the planet's temperature. If a planet is too hot, or too small, some of the molecules of atmospheric gases will always be moving faster than escape velocity. Unless they have a scattering collision with some other, slower, molecule, they will escape the planet completely. And unless they are replaced, from the interior or in some other way, the planet will at last lose its atmosphere.

A body as cool, big, and far from a star as Jupiter (escape velocity 60 kms/sec) or Saturn (escape velocity 36 kms/sec) is from the Sun will hold onto its atmosphere indefinitely. A body as small and hot as Mercury (escape velocity 4 kms/sec) or as small as Ceres (escape velocity 0.46 kms/sec) has no chance. Any atmosphere will vanish over time.

The surface gravity of a planet, g (or gee), a quantity with which we are more personally familiar, depends on exactly the same variables. We have $g=GM/r^2$, where M, r, and G are the same as before. For the case of Earth, we find $g=9.80$ m/sec^2.

In the past few years, the existence of planets around other stars has changed from optimistic guess to fairly confident reality. TABLE 7.5 (p. 190) gives a list of some of them, all admittedly based on evidence that is, if not weak, at least indirect. The list is representative rather than complete, because the number is growing fast. A new planet is added every month or two. We have not yet actually *seen* a planet around another star, even though every planet on the list is big, Jupiter's size or more.

That should not be taken to mean that most planets in the universe are massive. It merely shows that our detection methods can find only big planets. Possibly there are other, smaller planets in every system where a Jupiter-sized giant has been discovered.

Two planets in TABLE 7.5 are more than five times the mass of Jupiter. They are so big that these worlds are candidate "brown dwarf" stars, glowing dimly with their own heat. It is also disconcerting to see massive planets orbiting so close to their primary stars. In the case of 51 Pegasi and 55 Cancri, we have planets at least half the size of Jupiter, and perhaps a good deal bigger, orbiting only seven and sixteen million kilometers out from their sun. A planet of that size and in that position in our own solar system would have profound effects on Earth and the other inner planets.

If we cannot actually see a planet, how can we possibly know that they exist? There are two methods. First, it is not accurate to say that a planet orbits a star. The two bodies orbit around their common center of mass. That means, if the planet's orbit lies at right angles to the direction of the star as seen from Earth, the star's apparent position in the sky will show a variation over the period of the planetary year. That change will be tiny, but if the planet is large, the movement of the star may be big enough to measure.

The other, and so far more successful, method of detection also relies on the fact that the star and planet orbit around their common center of gravity, but in this case we look for a periodic shift in the wavelength of the light that we receive. When the star is approaching us because the planet is moving away from us, the light will be shifted toward the blue. When the star is moving away from us because the planet is approaching us, the star's light will be shifted toward the red. The tiny difference between these two cases allows us, from the wavelength changes in the star's light, to infer the existence of a planet in orbit around it.

Since both methods of detection depend for their success on the planet's mass being an appreciable fraction of the star's mass, it is no surprise that we are able to detect only the existence of massive planets, Jupiter-sized or bigger. The size distribution of planets around other stars remains an open question. Will we ultimately find a continuum, everything from small, Mercury-sized planets on up to planets able to sustain their own fusion reactions and thus to multiple star systems? Or are there major gaps in sizes, as we find in our own solar system between the inner and outer planets?

Are all stars candidates for planets that might support life? They are not, and we can narrow the search process. First, as noted in Chapter 3, massive stars burn their nuclear fuel much faster than small ones. A star ten times the mass of the sun will consume its substance several thousand times as rapidly. As a result, instead of continuing to shine as Sol will, more or less unchanged for over five billion years, our massive star will find its fuel exhausted in just a few million years. Its end, as we saw in Chapter 4, is cataclysmic. No planet could survive the explosion of its primary as a supernova.

The chance that native life, still less intelligence, might be wiped out in such a stellar conflagration is negligible. It would not have had time to develop. We do not know how long life took to establish itself on Earth, but it was surely longer than a few million years. The solar system was a turbulent place four and a half billion years ago, and Earth did not have a surface suitable to support life for at least the first few hundred million years. A planet orbiting a massive star would be gone before its crust had solidified.

Recall our horrible example from Chapter 1. The home world of the aliens orbited Rigel. But Rigel is a supergiant star, with a mass as much as 50 solar masses. It runs through its stable phase so fast that alien intelligence would have no time to develop. Add that to the list of story problems that need fixing.

We must deal with one other obstacle to the formation of planets suitable for life. The Sun is a star, and when we speak of, for example, Sirius or Rigel, we tend to think of them as single stars also. However, double and triple star systems are very common. Alpha Centauri, the nearest star to us, is actually three stars, labeled Alpha Centauri A, Alpha Centauri B, and Proxima Centauri ("proxima," meaning "close," refers to the star's distance from *us*, not from its companions; it is a tenth of a light-year away from the A and B components, and has an orbital period of at least half a million years). Since

Proxima is small and dim, as seen from a planet circling Alpha Centauri A or B it would not be among their top thirty bright stars.

In the same way, Sirius is two stars, Sirius A and Sirius B. The second is sometimes called the "dark companion," not because it is really dark, but because it is small and condensed. Its existence was deduced by Bessel in 1844, from observations he had made of the perturbation of the brighter Sirius A. However, no one saw the companion until Alvan Clark observed it in 1862. That only added to the mystery, because although calculations showed that Sirius B had to be about as massive as the Sun, it shone only one four-hundredth as bright. Sirius B is a white dwarf star, the first one discovered, and its average density is several tons per cubic inch. Finally, Rigel also has a companion—and the companion itself seems to be a binary star.

The relevance of all this to planets suitable for life is defined by celestial mechanics. When we have a star like our Sun, planetary orbits around it tend to be stable over long periods of time. The Earth has varied little in its distance from Sol, and hence in the amount of solar heating, during its whole lifetime. The mathematical description of the motion of the Earth around the Sun is provided by the "two-body problem," solved by Isaac Newton in the late seventeenth century. Perturbation effects of other planets, particularly Jupiter, were included by later workers such as Laplace, and confirmed the stability of the Earth's orbit.

When two or more stars are in one stellar system, however, the relevant mathematical problem for the motion of a planet is termed the "N-body problem." The formal exact solution has never been found, but approximate solutions can be obtained in any particular case, using computers. When this is done, the fate of a planet in an N-body system of multiple stars is found to be very different from the stable orbits of our own solar system. Orbits are far more chaotic. Close encounters of a planet with one or other of the primary stars will take place, distances vary wildly over time, and in extreme cases a combination of gravitational forces can eject the planet totally from the stellar system.

Even if the planet does not suffer such a fate, it moves through various extreme situations, now close to a star and baking in radiation, now far away in the freezing dark. This is, so far as we know, not a promising environment for the development of life.

There are two ways for a storyteller to avoid these problems. One is to be so blissfully ignorant of basic astronomy and astrophysics that you see no problem putting life and intelligence any place that you choose, and you hope for equal ignorance on the part of the reader. If you have come this far with me, you will know that I do not approve of such an approach.

The other way is to choose a star without companions, of a stellar type close to our own Sun. Suitable candidates that are also our stellar neighbors include Epsilon Eridani, at 11 light-years, and Tau Ceti, at 12 light-years. No one knows if either star has planets, though Epsilon Eridani has a ring of dust particles which is considered a promising sign. You are free to give either of these stars a world with the size and chemistry of Earth, and explain to the reader that this is the case.

You will then not have the chore of building a plausible world, and you will be safe from criticism. But as Hal Clement, justly famous for designing and explaining exotic worlds, says, "Where's the fun in that?"

TABLE 7.1. The Moons of Jupiter.

Physical properties

Name	Mass (10^{20} kg)	Radius (kms)	Density	Albedo
Galilean Satellites				
Io	893	1,821	3.530	0.61
Europa	480	1,565	2.990	0.64
Ganymede	1,482	2,634	1.940	0.42
Callisto	1,076	2,403	1.851	0.20
Lesser Satellites				
Metis		20±10		0.05
Adrastea		10±10		0.05
Amalthea		131×73×67		0.05
Thebe		50±10		0.05
Leda		5		
Himalia		85±10		
Lysithea		12		
Elara		40±10		
Ananke		10		
Carme		15		
Pasiphae		18		
Sinope		14		

Orbital parameters

Name	Semimajor axis (1000's kms)	Period* (days)	Inclination (degrees)	Eccentricity
Galilean satellites				
Io	422	1.769	0.040	0.041
Europa	671	3.552	0.470	0.0101
Ganymede	1,070	7.155	0.195	0.0015
Callisto	1,883	16.689	0.281	0.007
Lesser Satellites				
Metis	128	0.297	0	0.041
Adrastea	129	0.298	0	0
Amalthea	181	0.498	0.40	0.003
Thebe	222	0.675	0.8	0.0015
Leda	11,094	238.72	27	0.163
Himalia	11,480	250.56	28	0.163
Lysithea	11,720	259.22	29	0.107

Elara	11,737	259.65	28	0.207
Ananke	21,200	631R	147	0.169
Carme	22,600	692R	163	0.207
Pasiphae	23,500	735R	148	0.378
Sinope	23,700	758R	153	0.275

* The symbol R after the period indicates that the moon is in retrograde motion; i.e., it orbits in the opposite direction to Jupiter's rotation on its axis.

TABLE 7.2 The Moons of Saturn

Physical properties

Name	Mass (10^{20} kg)	Radius (km)	Density	Albedo
Mimas	0.38	198.8	1.140	0.5
Enceladus	0.73	249.1	1.120	1.0
Tethys	6.22	529.9	1.000	0.9
Dione	10.52	560	1.440	0.7
Rhea	23.10	764	1.240	0.7
Titan	1,345.50	2,575	1.881	0.21
Hyperion	185×140×113		0.19–0.25	
Iapetus	15.9	718	1.020	0.05–0.5
Phoebe 1	15x110x105			0.06

Lesser Satellites

Name	Mass	Radius	Density	Albedo
Pan		10		0.5
Atlas		18.5×17.2×13.5		0.9
Prometheus	0.0014	74×50×34	0.270	0.6
Pandora	0.0013	55×44×31	0.420	0.9
Epimetheus	0.0055	69×55×55	0.630	0.8
Janus	0.0198			
		99.3×95.6×75.6	0.650	0.8
Calypso		15×8×8		0.6
Telesto		15×12.5×7.5		0.5
Helene		16		0.7

Orbital parameters

Name	Semimajor axis (1000's kms)	Period* (days)	Inclination (degrees)	Eccentricity
Mimas	185.5	0.942	1.53	0.0202
Enceladus	238.0	1.370	0.02	0.0045
Tethys	294.7	1.888	1.09	0.0000
Dione	377.4	2.737	0.02	0.0022
Rhea	527.0	4.518	0.35	0.001
Titan	1,221.9	15.945	0.33	0.0292
Hyperion	1,481.1	21.277	0.43	0.1042
Iapetus	3,561.3	79.330	7.52	0.0283
Phoebe	12,952	550.48R	175.3	0.163

Lesser Satellites

Name	Semimajor axis (1000's kms)	Period* (days)	Inclination (degrees)	Eccentricity
Pan		133.6		0.575
Atlas	137.6	0.602	0	0
Prometheus	139.4	0.613	0.0	0.0024
Pandora	141.7	0.629	0.0	0.0042
Epimetheus	151.4	0.695	0.34	0.009
Janus	151.5	0.695	0.14	0.007
Calypso	294.7	1.888	0	0
Telesto	294.7	1.888	0	0
Helene	377.4	2.737	0.2	0.005

* The symbol R after the period indicates that the moon is in retrograde motion.

TABLE 7.3 The Moons of Uranus

Physical properties

Name	Mass (10^{20} kg)	Radius (km)	Density	Albedo
Miranda	0.659	240×234×233	1.200	0.27
Ariel	13.53	581×578×578	1.670	0.34
Umbriel	11.72	584.7	1.400	0.18
Titania	35.27	788.9	1.710	0.27
Oberon	30.14	761.4	1.630	0.24

Name	Mass (10²⁰ kg)	Radius (km)	Density	Albedo

Name	Mass (10^{20} kg)	Radius (km)	Density	Albedo
Lesser Satellites				
Cordelia		1		0.07
Ophelia		16		0.07
Bianca		22		0.07
Cressida		33		0.07
Desdemona		29		0.07
Juliet		42		0.07
Portia		55		0.07
Rosalind		29		0.07
Belinda		34		0.07
Puck		77		0.07

Orbital parameters

Name	Semimajor axis (1000's kms)	Period (days)	Inclination (degrees)	Eccentricity
Miranda	129.8	1.413	4.22	0.0027
Ariel	191.2	2.520	0.31	0.0034
Umbriel	266.0	4.144	0.36	0.0050
Titania	435.8	8.706	0.10	0.0022
Oberon	582.6	13.463	0.10	0.0008
Lesser Satellites				
Cordelia	49.75	0.335	0.1	0.000
Ophelia	53.76	0.376	0.1	0.010
Bianca	59.17	0.435	0.2	0.001
Cressida	61.78	0.464	0.0	0.000
Desdemona	62.66	0.474	0.2	0.000
Juliet	64.36	0.493	0.1	0.001
Portia	66.10	0.513	0.1	0.000
Rosalind	69.93	0.558	0.3	0.000
Belinda	75.26	0.624	0.0	0.000
Puck	86.00	0.762	0.3	0.000

TABLE 7.4 The Moons of Neptune.

Physical properties

Name	Mass (10²⁰ kg)	Radius (km)	Density	Albedo
Naiad		29		0.06
Thalassa		40		0.06
Despina		74		0.06
Galatea		79		0.06
Larissa		104×89		0.06
Proteus		218×208×201		0.06
Triton	214.7	1,352.6	2.054	0.7
Nereid		170		0.2

Orbital parameters

Name	Semimajor axis (1000's kms)	Period* (days)	Inclination (degrees)	Eccentricity
Naiad	48.23	0.29	4.74	0.00
Thalassa	50.08	0.31	0.21	0.00
Despina	52.53	0.33	0.07	0.00
Galatea	61.95	0.43	0.05	0.00
Larissa	73.55	0.56	0.20	0.00
Proteus	117.65	1.12	0.55	0.00
Triton	354.76	5.88R	156.83	
Nereid	5,513.4	360.14	7.23	0.75

* The symbol R after the period indicates that the moon is in retrograde motion.

TABLE 7.5
Planets of other stars.

Star	Distance of planet from star (Earth to Sun=1)	Minimum mass (Jupiter=1)	Orbit Period (days)
51 Pegasi	0.05	0.5	4.3
47 Ursae Majoris	2.1	2.4	1,103
70 Virginis	variable	6.6	117
55 Cancri	0.11	0.8	14.76
HD 114762	variable	10.0	84
Tau Bootis	0.0047	3.7	3.3
Upsilon Andromedae	0.054	0.6	4.61
Lalande 21185	2.2	0.9	5.8 (yrs)
HD 210277	1.15	1.36	1.2 (yrs)

CHAPTER 8
Spaceflight

8.1 The ways to space. The whole universe beyond Earth sits ready and waiting as our story setting. There is only one problem: How do we get there?

We will suggest a number of ways to move to and around in space. After a brief summary of each, we will push every one to its limit (and perhaps a little beyond).

The dozens of different systems for moving to and in space can be divided into three main types:

Category A, Rocket Spaceships. These achieve their motion via the expulsion of material that they carry along with them. Usually, but not always, the energy to expel the reaction mass comes from that reaction mass itself, by burning or through nuclear reactions. As an alternative, the energy to move reaction mass at high speed comes from another energy source, either on the rocket or elsewhere.

Category B, Rocketless Spaceships, which do not carry their own reaction mass. These ships must derive their motive force from some external agent.

Within each of the two major divisions we will find considerable variation. Category A includes:

- Chemical rockets.
- Mass drivers.
- Ion rockets.
- Nuclear reactor rockets.
- Pulsed fission rockets.

- Pulsed fusion rockets.
- Antimatter rockets.
- Photon rockets.

Category B includes:

- Gravity swingbys.
- Solar sails.
- Laser beam propulsion.
- The Bussard ramjet.

We will also mention a trio of hybrid systems:

- Laser-powered rockets.
- The Ram Augmented Interstellar Rocket (RAIR).
- The vacuum energy drive.

Finally, in a Category C we examine special devices able to take people and cargo into orbit without using rockets. Category C includes:

- Beanstalks.
- Dynamic beanstalks.
- Space fountains.
- Launch loops.

As we will see, some of these are suited only for in-space operations, while others are a natural choice for launch operations.

8.2 Rocket spaceships. *A chemical propulsion* system is just a fancy term for generating propulsion by ordinary burning (the burning takes place so rapidly and violently that we may prefer to think of it as a controlled explosion). This is a tried-and-tested standby, and every ounce of material launched into space today has been done using chemical rockets; yet in some ways, the rocket looks like the worst choice of all.

To see why, imagine that you have developed a wonderful new form of rocket that provides a significant thrust for many hours, or days, or even weeks, at the cost of very little fuel (fuel used by a rocket is termed "reaction mass," since the rocket is propelled forward as a reaction to the expelled fuel traveling backward).

In science fiction, and also in actual space travel, an acceleration equal to that produced by gravity on the surface of the Earth is called one *gee*. As we saw in Chapter 7, this is about 9.8 m/sec^2. Accelerations are then specified in multiples of this. For example, three gees, during the Shuttle's ascent to orbit, means that the astronauts will experience an acceleration of 29.4 m/sec^2, and no further explanation needs to be given. We will use the same convention.

Suppose, then, that the thrust of our new rocket engine is enough to generate an acceleration of half a gee. As we will see when we consider ways of moving around once we are in space, a half gee acceleration provides easy access to the whole solar system—once we have managed to get away from Earth.

We place our new rocket upright on the launch pad and switch on the engine. Reaction mass is expelled downward to provide an upward thrust. What happens next?

Absolutely nothing. When the total upward thrust is less than the weight of the rocket, the whole thing will simply sit there. Earth's gravity provides an effective "downward thrust" equal to the rocket's total weight, and unless the upward thrust provided by the propellant's ejection exceeds that weight, the rocket will not move one inch. We can fire our engine for hours or weeks or years, but we will not achieve any movement at all. (Even trained engineers can sometimes miss this basic point. In 1995 I received a proposal from a Canadian engineer for a launch system using an ion engine that could produce an acceleration of only a tiny fraction of a gee.)

Things are slightly better when the thrust of the rocket's engines is a little bigger than the total weight. Suppose that the thrust is 1.01 times the initial weight. The rocket will move upward, but agonizingly slowly. At first, the acceleration will only be one hundredth of a gee. The rocket will accelerate faster as it ascends, since it no longer lifts the weight of fuel already expelled. But it will still provide a puny acceleration. You can get to orbit that way, but it will take a long time. And for all that time, while you move slowly upward, your rocket is wasting thrust. Almost all the fuel being expelled as reaction mass is simply going to counteract the downward acceleration provided by the Earth itself.

Now it is clear why astronauts are trained to accept high accelerations. The faster that the rocket can burn its fuel, the higher the thrust will be, the higher the useful thrust (more than gravity's pull) will be, and the less the fuel wasted in reaching orbit. Once we are in orbit, fuel is no longer needed to fight Earth's gravity. Hence the old maxim of spaceflight, *once you are in orbit you are halfway to anywhere.*

Notice the basic difference between flight *to space*, and flight *in space*. Any rocket engine that provides less than one gee of acceleration cannot take us up to orbit. Once we are in space, however, and in some orbit, there is no lower limit below which an acceleration is useful. Any acceleration, no matter how small, can be used to transfer between any two orbits, though it may take a while to accomplish the move.

You may object to one of the assumptions made in this analysis. Why place the rocket vertically? Suppose instead that we placed the rocket horizontally, on a long, smooth railroad track. Then any acceleration, no matter how small, will speed up the rocket, since horizontal motion does not fight against Earth's gravity. If we keep increasing in speed, eventually we will be moving so fast that the rocket would have to be held down on the track, otherwise centrifugal force would make it rise. If we could reach a speed of eight kilometers a second before releasing from the track, the rocket would be going fast enough to take it up to orbit.

This is good science and good engineering—but not for launch from Earth. The atmosphere is the spoiler, making the method impractical because of air resistance and heating. For the Moon, however, with its negligible atmosphere, the method will be perfectly fine. It was suggested long ago as a good lunar launch technique.

First, however, we have to be on the Moon, or out in space. That takes us back to where we started. We conclude as many have concluded: beginning the exploration of space from the surface of the Earth may be our only option, but given a choice we would not start from here.

Unfortunately, we *are* here. Back to our launch problem, leaving from the surface of the Earth.

8.3 Measures of performance. We need some way to evaluate rocket propulsion systems, so that we can say, "Of these types of rocket-propelled systems, Type A has greater potential than Type B."

With chemical fuels such as kerosene or liquid hydrogen, it is natural to look for guidance from the way in which we compare fuels here on Earth. An easy general measure happens to be available: the number of kilocalories produced when we burn a gram of the fuel. For instance, good coal will yield about 7 kilocalories a gram, gasoline about 11.5 kilocalories a gram. Based on that measure alone, we would expect to prefer gasoline to coal as an energy source, and indeed, the coal-powered spaceship does not feature largely in science fiction.

However, for rocket propulsion the heat generated by burning the fuel is not quite the measure that we want. The right variable is the *specific impulse* of a given type of fuel, and it measures the *thrust* that the fuel can generate. Specific impulse, usually written as SI, is the length of time that one pound of fuel can produce a thrust of one pound weight. SI is normally measured in seconds. Since weight depends on the value of surface gravity, and since surface gravity depends where you are on Earth (it is more at the poles than at the equator) this may seem like a rather poor definition. It came into use in the 1920s and 1930s, when people doing practical experiments with rockets found it a lot easier to measure the force that a rocket engine was developing on a stand than to measure the speed of the expelled gases that formed the rocket's exhaust. That speed is a better measure, and it is termed the *effective jet velocity*, or EJV. We say effective velocity rather than actual velocity, because if the reaction mass is not expelled in the desired direction (opposite to the spacecraft's motion) the EJV will be reduced. Thus EJV measures both the potential thrust of a fuel, and also the efficiency of engine design.

We will use mainly EJV. However, SI is still a widely-used measure for comparing different rocket fuels, so it is worth knowing about. To convert from SI values (in seconds) to EJV (in kilometers per second), simply multiply by 0.0098.

Naturally, neither SI nor EJV is a useful measure in propulsion systems that do not employ reaction mass. However, they are supremely important factors in near-future practical spaceflight, because the ratio of final spacecraft mass (payload) to initial mass (payload plus fuel) depends *exponentially* on the EJV.

Explicitly, the relationship is $MI/MP = e^{(V/EJV)}$, where MI=initial total mass of spacecraft plus fuel, MP=final payload mass, and V=final spacecraft velocity. This is often termed the Fundamental Equation of Rocketry. It is true only at nonrelativistic speeds, so some day we may have to change our definition of "Fundamental."

To see the importance of this equation, suppose that a mission has been designed in which the initial mass of payload plus fuel is 10,000 times the final payload. That is a prohibitively high value for most missions, and the design is useless. But if the EJV of the mission could somehow be doubled, the initial payload-plus-fuel mass would become only 100 (the square root of 10,000) times the final payload. And if it could somehow be doubled again, the payload would increase to one-tenth (the square root of 1/100) of the initial total mass. The secret to high-performance missions lies in high values of the EJV.

Very good. But what kind of values of the EJV might we expect in the best rocket system?

The chemical rockets that we can make with present technology, using a liquid hydrogen/liquid oxygen (LOX) mix, produce an EJV rather more than 4 kilometers per second. LOX plus kerosene is less good, with an EJV of about 2.6. Potassium perchlorate plus a petroleum product (a solid fuel rocket) has an EJV of about 2. Liquid hydrogen with liquid fluorine—a tricky mixture to handle, with unpleasant combustion products—has an EJV as high as 4.5. That is probably the limit for today's chemical fuel rockets. To do better than this we would have to go to such exotic fuels as monomolecular hydrogen, which is highly unstable and dangerous.

As we noted in Chapter 5, the maximum performance for chemical fuels occurs when the energy released is perfectly converted to kinetic energy. The theoretical values obtained there were 5.6 kms/sec for the $H-O^{16}$ mix, and 5.95 kms/sec for $H-O^{14}$. If you write a story and your chemical-fuel rocket has an EJV of 50 (i.e. an SI of 5,100), you'll have to provide a pretty good explanation of how you did it; otherwise, you are writing not science fiction but pure fantasy.

Luckily for the writer, the chemical rocket is not the only one available; it merely happens to be the only type used so far for launches. That promises to be true for some time in the future. Even NASA's "new" development, the single-stage-to-orbit reusable rocket, will be a chemical rocket.

8.4 Mass drivers. The mass driver consists of a long helical spiral of wire with a hollow center (a *solenoid*). Pulsed magnetic fields are used to propel each payload along the solenoid, accelerating it until it reaches the end of the solenoid and flies off at high speed.

Mass drivers are usually thought of as launch devices, throwing payloads to space using electromagnetic forces. However, suppose that we invert our thinking. A mass driver in free space will itself be given an equal push by the material that is expelled (Newton's Third Law: Action and reaction are equal and opposite). If we regard the expelled material as reaction mass, then the long solenoid itself is part of the spacecraft, and it will be driven along in space with the rest of the payload.

Practical tests suggest that ejecting a series of small objects using the mass driver can give an EJV to the mass driver itself of up to 8 kms/second. This is almost double the EJV that can be achieved with chemical rockets; however, note that the energy to power the mass driver must be provided externally, for instance as electricity generated using nuclear or solar power. The mass of such power-generation equipment will diminish the mass driver's performance as a propulsion system. Mass drivers do not offer a solution to the problem of reaching orbit from the surface of the

Earth. In addition, solar power is fine, close to the Sun, but it would be a major problem out at the edge of the solar system. Available solar energy falls off as the inverse square of the distance from the Sun. We will encounter the same problem later, with other systems.

The good news is that working mass drivers have been built. They are not just theoretical ideas.

8.5 Ion rockets are similar in a sense to mass drivers, in that the reaction mass is accelerated electromagnetically, and then expelled. In this case, however, the reaction mass consists of charged atoms or molecules, and the acceleration is provided by an electric field. The technique is the same as that used in the linear accelerators employed in particle physics work here on Earth. Very large linear accelerators, miles in length, have already been built; for example, the Stanford Linear Accelerator (SLAC) has an acceleration chamber two miles long.

SLAC is powered using conventional electric supplies. For use in space, the power supply for ion rockets can be solar or nuclear (or externally provided; see the discussion of laser power later in this chapter). As was the case with mass drivers, provision of that power supply must diminish system performance.

Prototype ion rockets have been flown in space. They offer a drive that can be operated for long periods of time, and thus they are attractive for long missions. Practical tests suggest that they can produce an EJV of up to 70 kms/second, far higher than the EJV of either chemical rockets or mass drivers. However, because the onboard equipment to produce the ion beam is bulky, these are low-thrust devices providing accelerations of a few micro-gees. In order to achieve final velocities of many kilometers per second, ion rockets must be operated for long periods of time. They are not launch devices.

8.6 Nuclear reactor rockets use a nuclear reactor to heat the reaction mass, which is then funneled to expel itself at high temperatures and at high velocities.

Systems with a solid core to the reactor achieve

working temperatures up to about 2,500°C, and an EJV of up to 9.5 kms/second. Experimental versions were built in the early 1970's. Work on the most developed form, known as NERVA, was abandoned in 1973, because of concern about spaceborne nuclear reactors. A solid core reactor rocket with hydrogen as reaction mass has an EJV more than double the best chemical fuel rocket, but the nuclear power plant itself has substantial mass. This reduces the acceleration to less than a tenth of what can be achieved with chemical fuels.

A liquid core reactor potentially offers higher performance, with a working temperature of up to 5,000°C and an EJV of up to 25 kms/second. Gaseous core reactors can do even better, operating up to 20,000°C and producing an EJV of 65 kms/second. However, such nuclear reactor rockets have never been produced, so any statements on capability are subject to question and practical proof.

I believe we could go to orbit with a liquid core nuclear-powered rocket, safely and more efficiently than with a chemical rocket. However, I think it will be some time before we are allowed to. The suspicion of nuclear launch—or, indeed, all things nuclear—is too strong.

8.7 Pulsed fission rockets form the first of the "advanced systems" that we will consider; advanced, in the sense that we have never built one, and doing so might lead to all sorts of technological headaches; and also advanced in the sense that such rockets, if built, could take us all over the solar system—and out of it.

The idea for the pulsed fission rocket may sound both primitive and alarming. A series of atomic bombs (first design) or hydrogen bombs (later designs) are exploded behind the spacecraft, which is protected by a massive "pusher plate." This plate serves both to absorb the momentum provided by the explosions, and also to shield the payload from the radioactive blasts.

The pulsed fission rocket was proposed by Stanislaw Ulam in 1955. The idea, later known as Project Orion, appeared practical and could have been built. However, the effort was abandoned in 1965, a casualty of the

1963 Nuclear Test Ban Treaty. Project Orion called for full-scale atomic explosions, and the treaty made it impossible to test the idea. The EJV is excellent, up to 100 kms/second, but the mass of the pusher plate may limit practical accelerations to a few centimeters/sec^2 (less than a hundredth of a gee). This is no good for a launch system, but it will achieve respectable velocities over long periods. An acceleration of 1 cm/sec^2 (just over a thousandth of a gee) for one year produces an end speed of 310 kms/second. Note that, even at this speed, Neptune is still more than six months travel time away.

8.8 Pulsed fusion. The pulsed fission rocket of Project Orion has two big disadvantages. First, the nuclear explosions are full-scale nuclear blasts, each one equivalent in energy release to thousands or even millions of tons of conventional explosives. Second, the massive "pusher plate" is useful as a protection against the blasts and as an absorber of momentum, but it greatly decreases the acceleration of the ship and the system efficiency.

The pulsed *fusion* rocket potentially overcomes both these problems. Each fusion explosion can be a small one, involving only a gram or so of matter. The fusion process is initiated by a high-intensity laser or a relativistic electron beam focused on small spheres of nuclear fuel. The resulting inward-traveling shock wave creates temperatures and pressures at which fusion can occur. If the right nuclear fuels are used, all the fusion products can be charged particles. Their subsequent movement can therefore be controlled with electromagnetic fields, so that they do not impinge on the payload or the walls of the drive chamber.

An analysis of a pulsed fusion rocket mission was performed in the late 1970s by the British Interplanetary Society. Known as Project Daedalus, it was a design for a one-way trip to Barnard's Star, 5.9 light-years from the Sun. Small spheres of deuterium (D) and helium-3 (He^3) were used as fusion fuels. (Deuterium is "heavy" hydrogen, $_1$H^2, with a neutron as well as a proton in the nucleus; helium-3, $_2He^3$, is "light" helium, missing a neutron in its nucleus.)

The D-He^3 reaction yields as fusion products a helium nucleus and a proton, both of which carry electric charges and can thus be manipulated by magnetic fields. The estimated EJV for Project Daedalus was 10,000 kms/second, leading to a fifty-year travel time for the 5.9 light-year journey. The mass at launch from solar orbit was 50,000 tons, the final mass was 1,000 tons, and the terminal velocity for the spacecraft was one-eighth of the speed of light.

The design was a technical tour de force, but the complications and caveats are significant. First, controlled pellet fusion of the type envisaged has never been demonstrated. The D-He^3 fusion reaction in the fuel pellets proceeds rapidly only at extreme temperatures, and while other fusion reactions, such as deuterium-tritium, take place at a sixth of this temperature, they produce uncharged neutrons as fusion products and the direction of travel of these uncharged particles cannot easily be controlled.

Third, and perhaps the biggest problem of all, the nuclear fuels needed are not available. Deuterium is plentiful enough, at one part in 6,000 in ordinary hydrogen. But He^3 is very rare on Earth. The total U.S. supply is only a few thousand liters. The Daedalus design calls for 30,000 tons of the stuff, far more than could be found anywhere on Earth. The only place in the solar system where He^3 exists in enormous quantities is in the atmospheres of the gas-giant planets, Jupiter and Saturn and Uranus and Neptune.

Project Daedalus proposed the use of a complicated twenty-year mining operation in the atmosphere of Jupiter, to be conducted by automated factories floating in the Jovian atmosphere. The construction of the spacecraft itself would be carried out near Jupiter. I took over their method in my novel *Cold as Ice*, but I assumed that the moons of Jupiter had already been colonized by humans. Access—and management oversight—was easier, and in fact the necessary helium mining formed only a minor element of the book.

8.9 Antimatter rockets. To every particle in nature there corresponds an antiparticle. Matter constructed from these antiparticles is termed antimatter, or mirrormatter. For example, antihydrogen consists of a positron moving about an antiproton, whereas normal hydrogen is an electron moving about a proton.

When matter and antimatter meet, they annihilate each other. They therefore represent a vast source of potential energy.

If electrons and positrons meet, the result is high-energy gamma rays, and no particles. If protons and antiprotons meet, the result is an average of three charged pions and two uncharged pions, with the charged pions carrying 60 percent of the total energy. Neutral pions decay to form high-energy gamma rays in less than a thousand trillionth of a second. Charged pions last a lot longer, relatively speaking, decaying to the elementary particle known as a muon in 26 nanoseconds. Muons decay in their turn to electrons and neutrinos, lasting on average 2.2 microseconds before they do so.

These are short times, but relativity helps here. The charged pions created in this process are traveling fast, at over ninety percent of the speed of light, and thus the effect of relativistic time dilation is to increase their lifetime from 26 nanoseconds to 70 nanoseconds. This is more than long enough to control the movement of the charged pions with magnetic fields. Similarly, the rapidly-moving muons that appear as decay products last on average 6.2 microseconds rather than 2.2 microseconds, before they in turn decay. They too can be controlled through the use of magnetic fields.

Antimatter is a highly concentrated method of storing energy. The total energy produced by a milligram of antimatter when it meets and annihilates a milligram of ordinary matter is equal to that of twenty tons of liquid hydrogen/LOX fuel. It is therefore ideal for use on interstellar missions, where energy per unit weight is of paramount importance in fuels.

The most economical way of using such a potent fuel is not to take it "neat," but to dilute the antimatter with a large amount of ordinary matter. Matter/antimatter annihilation then serves to heat up ordinary matter, which is expelled as reaction mass. In this case, both the high-energy gamma rays and the pions serve to heat the reaction mass; and by choosing the antimatter/matter ratio, many different missions can be served with a single engine design. A highly dilute matter/antimatter engine also has excellent potential for interplanetary missions.

Given all these useful properties of antimatter, why are we waiting? Well, one question remains: How do we get our hands on some of this stuff?

That leads us to one of the major mysteries of physics and cosmology. There is as much reason for antimatter to exist as for ordinary matter to exist. Logically, the universe should contain equal amounts of each. In practice, however, antimatter is very rarely found in nature. Positrons and antiprotons occur occasionally in cosmic rays, but if we discount the highly unlikely possibility that some of the remote galaxies are all antimatter, then the universe is ordinary matter to an overwhelming extent.

One product of the recent inflationary models of the early universe is a possible explanation of the reason why there is so little antimatter. This, however, is of little use to us. We need antimatter *now*, and in substantial quantities, if we are to use matter-antimatter annihilation to take us to the stars.

Since antimatter is not available in nature, we will have to make our own. And this is possible. One by-product of the big particle accelerators at Fermilab in Illinois, at IHEP in Novosibirsk in the Soviet Union, and at CERN in Switzerland, is a supply of antiprotons and positrons. The antiprotons can be captured, slowed down, and stored in magnetic storage rings. Anti-hydrogen can be produced, by allowing the antiprotons to capture positrons. Antimatter can be stored in electromagnetic ion traps, and safely transported in such containers.

We are not talking about large quantities of antimatter with today's production methods. Storage rings have held up to a trillion antiprotons, but that is still a very small mass (about a trillionth of a gram). And antimatter takes a lot of energy to produce. The energy we will get from the antimatter will not be more than 1/10,000th of the energy that we put into making it. However, the concentrated energy of the end product makes this a unique fuel for propulsion.

The EJV of a matter/antimatter engine depends on the matter-to-antimatter ratio, and it can be selected to match the needs of particular missions. However, for interstellar travel we can safely assume that we want the biggest value of the EJV that we can get. This will occur when we use a 1:1 ratio of matter to antimatter, and direct the charged pions (and their decay products, the muons) with magnetic control of their final emission direction. Since the charged pions contain 60 percent of the proton-antiproton annihilation energy, and since the uncharged pions and the gamma rays will be emitted in all directions equally, we find the maximum EJV to be 180,000 kms/second. With such an EJV, and a ratio of initial mass to final mass of 3:1, the terminal velocity of the mission will be almost two-thirds of the speed of light. We are in a realm of velocities where relativistic effects have a big effect on shipboard travel times.

8.10 Photon rockets. This takes the matter-antimatter rocket to its ultimate form. It represents the final word in rocket spaceships that employ known physics.

If we could completely annihilate matter, so that it appeared as pure radiation (and was heading in the right direction, as a collimated beam), the EJV would be the speed of light, about 300,000 kilometers per second.

This is the highest EJV possible. It implies perfect magnetic control and redirection of all charged pions, plus the control of all uncharged pions and gamma rays and of all decay products such as electrons and neutrinos. Every particle produced in matter-antimatter annihilation ultimately decays to radiation, or to electrons and positrons that can then annihilate each other to give pure radiation. All this radiation must be emitted in a direction exactly opposite to the spacecraft's motion.

If the best chemical rocket with a fuel-to-payload ratio

of 10,000:1 could be replaced with a photon rocket, the mission would be 99.99 percent payload; the fuel would be a negligible part of the total mass. Having said that, we must also say that we have no idea how to make a photon rocket. It could exist, according to today's physics; but it is quite beyond today's technology.

8.11 Space travel without reaction mass. The central problem of the rocket spacecraft is easy to identify. For low to moderate EJV's (which we will define as less than 100 kms/second—a value that would make any of today's rocket engineers ecstatic) most of the reaction mass does not go to accelerate the payload. It goes to accelerate the rest of the fuel. This is particularly true in the early stages of the mission, when the rocket may be accelerating a thousand tons of fuel to deliver ten tons of payload. All systems carrying their reaction mass along with them suffer this enormous intrinsic disadvantage. It seems plausible, then, that systems which do not employ reaction mass at all may be the key to successful space travel. We now consider:
- Gravity swingbys.
- Solar sails.
- Laser beam propulsion.
- The Bussard ramjet.

Also, we will touch on three hybrid systems:
- Laser-powered rockets.
- The Ram Augmented Interstellar Rocket (RAIR).
- The vacuum energy drive.

8.12 Gravity swingbys. There is one form of velocity increase that needs neither onboard rockets nor an external propulsion source. In fact, it can hardly be called a propulsion system in the usual sense of the word. If a spacecraft flies close to a planet it can, under the right circumstances, obtain a velocity boost from the planet's gravitational field. This technique is used routinely in interplanetary missions. It was used to get the Galileo spacecraft to Jupiter, and to permit Pioneer 10 and 11 and Voyager 1 and 2 to escape the solar system. Jupiter, with a mass 318 times that of Earth, can give a velocity kick of up to 30 kms/second to a passing space-craft. So far as the spaceship is concerned, there will be no feeling of onboard acceleration as the speed increases. An observer on the ship experiences free fall, even while accelerating relative to the Sun.

If onboard fuel is available to produce a velocity change, another type of swingby can do even better. This involves a close approach to the Sun, rather than to one of the planets. The trick is to swoop in close to the solar surface and apply all available thrust near perihelion, the point of closest approach.

Suppose that your ship has a small velocity far from the Sun. Allow it to drop toward the Sun, so that it comes close enough almost to graze the solar surface. When it is at its closest, use your onboard fuel to give a 10 kms/second kick in speed; then your ship will move away and leave the solar system completely, with a terminal velocity far from the Sun of 110 kms/second.

The question that inevitably arises with such a boost at perihelion is, where did that "extra" energy come from? If the velocity boost had been given without swooping in close to the Sun, the ship would have left the solar system at 10 kms/second. Simply by arranging that the same boost be given near the Sun, the ship leaves at 110 kms/second. And yet the Sun seems to have done no work. The solar energy has not decreased at all. It sounds impossible, something for nothing.

The answer to this puzzle is a simple one, but it leaves many people worried. It is based on the fact that kinetic energy changes as the square of velocity, and the argument runs as follows: The Sun increases the speed of the spacecraft during its run towards the solar surface, so that our ship, at rest far from Sol, will be moving at 600 kms/second as it sweeps past the solar photosphere. The kinetic energy of a body with velocity V is $V^2/2$ per unit mass, so for an object moving at 600 kms/second, a 10 kms/second velocity boost increases the kinetic energy per unit mass by $(610^2-600^2)/2=6{,}050$ units. If the same velocity boost had been used to change the speed from 0 to 10 kms/second, the change in kinetic energy per unit mass would have been only 50 units.

Thus by applying our speed boost at the right moment, when the velocity is already high, we increase the energy change by a factor of $6,050/50 = 121$, which is equivalent to a factor of 11 (the square root of 121) in final speed. Our 10 kms/second boost has been transformed to a 110 kms/second boost.

All that the Sun has done to the spaceship is to change the speed relative to the Sun at which the velocity boost is applied. The fact that kinetic energy goes as the square of velocity does the rest.

If this still seems to be getting something for nothing, in a way it is. Certainly, no penalty is paid for the increased velocity—except for the possible danger of sweeping in so close to the Sun's surface. And the closer that one can come to the center of gravitational attraction when applying a velocity boost, the more gratifying the result.

Let us push the limits. One cannot go close to the Sun's center without hitting the solar surface, but an approach to within 20 kilometers of the center of a neutron star of solar mass would convert a 10 kms/second velocity boost provided at the right moment to a final departure speed from the neutron star of over 1,500 kms/second. An impressive gain, though the tidal forces derived from a gravitational field of over 10,000,000 gees might leave the ship's passengers a little the worse for wear.

Suppose one were to perform the swingby with a speed much greater than that obtained by falling from rest? Would the gain in velocity be greater? Unfortunately, it works the other way round. The gain in speed is maximum if you fall in with zero velocity from a long way away. In the case of Sol, the biggest boost you can obtain from your 10 kms/second velocity kick is an extra 100 kms/second. That's not fast enough to take us to Alpha Centauri in a hurry. A speed of 110 kms/second implies a travel time of 11,800 years.

8.13 Solar sails. If gravity swingbys of the Sun or Jupiter can't take us to the stars fast enough, can anything else? The Sun is a continuous source of a possible propulsive force, namely, solar radiation pressure. Why not build a large sail to accelerate a spacecraft by simple photon and emitted particle pressure?

We know from our own experience that sunlight pressure is a small force—we don't have to "lean into the sun" to stay upright. Thus a sail of large area will be needed, and since the pressure has to accelerate the sail as well as the payload, we must use a sail of very low mass per unit area.

The thinnest, lightest sail that we can probably make today is a hexagonal mesh with a mass of about 0.1 grams/square meter. Assuming that the payload masses much less than the sail itself, a ship would accelerate away from Earth orbit to interstellar regions at 0.01 gees.

This acceleration diminishes farther from the Sun, since radiation pressure per unit area falls off as the inverse square of the distance. Even so, a solar sail starting at 0.01 gees at Earth orbit will be out past Neptune in one year, 5 billion kilometers away from the Sun and traveling at 170 kms/second. Travel time to Alpha Centauri would be 7,500 years. Light pressure from the target star could be used to slow the sail in the second half of the flight.

8.14 Laser beam propulsion. If the acceleration of a solar sail did not decrease with distance from the Sun, the sail we considered in the last section would have traveled ten times as far in one year, and would be moving at 3,100 kms/second. This prompts the question, can we provide a constant force on a sail, and hence a constant acceleration, by somehow creating a tightly focused beam of radiation that does not fall off with distance?

Such a focused beam is provided by a laser, and this idea has been explored extensively by Robert Forward in both fact and fiction (see, for example, his *Flight of the Dragonfly*, aka *Rocheworld*; Forward, 1990). In his design, a laser beam is generated using the energy of a large solar power satellite near the orbit of Mercury. This is sent to a transmitter lens, hanging stationary out between Saturn and Uranus. This lens is of Fresnel ring type, 1,000 kilometers across, with a mass of 560,000 tons. It can send a laser beam 44 light-years without significant beam spreading, and a circular lightsail with a mass of 80,000 tons and a payload of 3,000 tons can be accelerated at that distance at 0.3 gees. That is enough to move the sail at half the speed of light in 1.6 years.

Forward also offers an ingenious way of stopping the sail at its destination. The circular sail is constructed in discrete rings, like an archery target. As the whole sail approaches its destination, one inner circle, 320 kilometers across and equal in area to one-tenth of the original sail, is separated from the outer ring. Reflected laser light from the outer ring serves to slow and halt the inner portion at the destination star, while the outer ring flies on past, still accelerating. When exploration of the target stellar system is complete, an inner part of the inner ring, 100 kilometers across and equal in area to one-tenth of the whole inner ring, is separated from the rest. This "bull's-eye" is now accelerated back towards the Sun, using reflected laser beam pressure from the outer part of the original inner ring. The travel time to Alpha Centauri, including slowing-down and stopping when we arrive, is 8.6 years (Earth time) and 7 years (shipboard time). Note that we have reached speeds where relativistic effects make a significant difference to perceived travel times.

Could we build such a ship, assuming an all-out worldwide effort?

Not yet. The physics is fine, but the engineering would defeat us. The power requirement of the laser is thousands of times greater than the total electrical production of all the nations on Earth. The space construction capability is also generations ahead of what can reasonably be projected for the next half century. We are not likely to go to the stars this way. Something better will surely come along before we are ready to do it. I feel this way about some other ideas, discussed later in this chapter.

8.15 The Bussard Ramjet. This is a concept introduced by Robert Bussard in 1960. It was employed in one of science fiction's classic tales of deep space and time, Poul Anderson's *Tau Zero* (Anderson, 1970).

In the Bussard ramjet, a "scoop" in front of the spaceship funnels interstellar matter into a long hollow cylinder that comprises a fusion reactor. The material collected by the scoop undergoes nuclear fusion, and the reaction products are emitted at high temperature and velocity from the end of the cylinder opposite to the scoop, to propel the spacecraft. The higher the ship's speed, the greater the rate of supply of fuel, and thus the greater the ship's acceleration. It is a wonderfully attractive idea, since it allows us to use reaction mass without carrying it with us. There is interstellar matter everywhere, even in the "emptiest" reaches of open space.

Now let us look at the "engineering details."

First, it will be necessary to fuse the fuel on the fly, rather than forcing it to accelerate until its speed matches the speed of the ship. Otherwise, the drag of the collected fuel will slow the ship's progress. Such a continuous fusion process calls for a very unusual reactor, long enough and operating at pressures and temperatures high enough to permit fusion while the collected interstellar matter is streaming through the chamber.

Second, interstellar matter is about two-thirds hydrogen, one-third helium, and negligible proportions of other elements. The fusion of helium is a complex process that calls for three helium nuclei to interact and form a carbon nucleus. Thus the principal fusion reaction of the Bussard ramjet will be proton-proton fusion. Such fusion is hindered by the charge of each proton, which repels them away from each other. Thus pressures and temperatures in the fusion chamber must be extremely high to overcome that mutual repulsion.

Third, there is only about one atom of interstellar matter in every cubic meter of space. Thus, the scoop will have to be many thousands of kilometers across if hydrogen is to be supplied in enough quantity to keep a fusion reaction going. It is impractical to construct a material scoop of such a size, so we will be looking at some form of magnetic fields.

Unfortunately, the hydrogen of interstellar space is mainly neutral hydrogen, i.e., a proton with an electron moving around it. Since we need a charged material in order to be able to collect it electromagnetically, some method must first be found to ionize the hydrogen. This can be done using lasers, beaming radiation at a carefully selected wavelength ahead of the ramjet. It is not clear that a laser can be built that requires less energy than is provided by the fusion process. It is also not clear that materials exist strong enough to permit construction of a magnetic scoop with the necessary field strengths.

The Bussard ramjet is a beautiful concept. Use it in stories by all means. However, I am skeptical that a working model will be built any time within the next couple of centuries, or perhaps ever.

8.16 Hybrids. For completeness, we will also mention three other systems. One has an onboard energy source and uses external reaction mass, the other two have onboard reaction mass and use external energy.

8.17 Laser-powered rockets. These rockets carry reaction mass, but that mass does not produce the energy for its own heating and acceleration. Instead, the energy is provided by a power laser, which can be a considerable distance from the target spaceship.

This concept was originally proposed by Arthur Kantrowitz as a technique for spacecraft launch. It is attractive for interplanetary missions, although for laser power to be available at interstellar distances it is necessary to build a massive in-space power laser system.

The requirement for onboard storage of reaction mass is also huge. Even when all of this has been done, the EJV does not exceed maybe 200 kms/second. This system sounds fine for launches, less good for in-space use. Although we never named it as such, this is what Jerry Pournelle and I used as the launch system in our novel *Higher Education* (Sheffield and Pournelle, 1996).

Note that laser power could be used equally well to provide the energy for other propulsion systems, such as the ion drive. This removes the bulky onboard equipment that otherwise severely limits ship acceleration.

8.18 Ram Augmented Interstellar Rocket (RAIR). The RAIR employs a Bussard ramscoop to collect interstellar matter. However, instead of fusing such matter as it flashes past the ship, in the RAIR an onboard fusion reactor is used to heat the collected hydrogen and helium, which then exits the RAIR cylinder at high speed.

Certainly, this eliminates one of the central problems of the Bussard ramjet—namely, that of fusing hydrogen quickly and efficiently. It also allows us to make use of interstellar helium. However, the other problems of the Bussard ramjet still exist. One little-mentioned problem with both the RAIR and the original Bussard ramjet is the need to reach a certain speed before the fusion process can begin, since below that speed there will not be enough material delivered to the fusion system. The acceleration to reach that minimum velocity is itself beyond today's capabilities.

8.19 The vacuum energy drive. The most powerful theories in physics today are quantum theory and the theories of special and general relativity. Unfortunately, those theories are not totally consistent with each other. If we calculate the energy associated with an absence of matter—the "vacuum state"—we do not, as common sense would suggest, get zero. Instead, quantum theory assigns a specific energy value to a vacuum.

In classical thinking, one could argue that the zero point of energy is arbitrary, so we could simply start measuring energies from the vacuum energy value. However, if we accept general relativity that option is denied to us. Energy, of any form, produces spacetime curvature, and we are therefore not allowed to redefine the origin of the energy scale. Once this is accepted, the energy of the vacuum cannot be talked out of existence. It is real, and when we calculate it we get a large positive value per unit volume.

How large?

Richard Feynman addressed the question of the vacuum energy value and computed an estimate for the equivalent mass per unit volume. The estimate came out as two billion tons per cubic centimeter. The energy in two billion tons of matter is more than enough to boil all Earth's oceans.

Is there any possibility that the vacuum energy could be tapped for useful purposes? Robert Forward has proposed a mechanism, based upon a real physical phenomenon known as the Casimir Effect. I think it would work, but the energy produced is small. The well-publicized mechanisms of others, such as Harold Puthoff, for extracting vacuum energy leave me totally unpersuaded.

Science fiction that admits it is science fiction is another matter. According to Arthur Clarke, I was the first person to employ the idea of the vacuum energy drive in fictional form, in the story "All the Colors of the Vacuum" (Sheffield, 1981). Clarke employed one in *The Songs of Distant Earth* (Clarke, 1986). Not surprisingly, there was a certain amount of hand-waving on both Clarke's part and mine as to how the vacuum energy drive was implemented. If the ship can obtain energy from the vacuum, and mass and energy are equivalent, why can't the ship get the reaction mass, too? How does the ship avoid being slowed when it takes on energy, which has an equivalent mass that is presumably at rest? If the vacuum energy is the energy of the ground state, to what new state does the vacuum go, after energy is extracted?

Good questions. Look on them as an opportunity. There must be good science-fictional answers to go with them.

8.20 Launch without rockets.

The launch of a rocket—any rocket—is certainly an impressive sight and sound. All that noise, all that energy, thousands of tons of fuel going up in smoke (literally) in a few minutes.

But does it have to be that way? Let us invoke a classical result from mathematics: *The work done carrying a test particle around a closed curve in a fixed potential field is zero.*

Around the Earth there is, to good accuracy, a fixed potential field. A spaceship that goes up to orbit and comes back down to the same place is following a closed curve. Conclusion: we ought to be able to send a test particle (such as a spacecraft, which on the scale of the whole Earth is no more than a particle) to orbit and back, without doing any work.

Let's do it, in several different ways.

8.21 The beanstalk. Suppose we have a space station in geostationary orbit, i.e. an equatorial orbit with period exactly 24 hours. A satellite in such an orbit hovers always over the same point on the Earth's equator. Such orbits are already occupied by communications satellites and some weather satellites.

Now suppose a strong loop of cable runs all the way down to the surface of Earth from the space station. The cable must be long as well as strong, since geostationary orbit is more than 35,000 kilometers above the surface. We defer the question as to how we install such a thing. (A *geostationary* satellite has a period of 24 hours, and hovers above a fixed point on the equator. A *geosynchronous* satellite simply has a period of 24 hours, but can be inclined to the equator and reach to any latitude.)

Attach a massive object (say, a new communications satellite) to the cable down on the surface. Operate an electric motor, winding the cable with the attached payload up to the station. We will have to do work to accomplish this, lifting the payload against the downward gravitational pull of the Earth. We do not, however, have to lift the cable, since the weight of the descending portion of the loop will exactly balance the weight of the ascending portion.

Also, suppose that we arrange things so that, at the same time as we raise the payload up from the surface, we lower an equal mass (say, an old, worn-out communications satellite) back down to the surface of the Earth. We will have to restrain that mass, to stop it from falling. We can use the force produced by the downward pull to drive a generator, which in turn provides the power to raise the payload. The only net energy needed is to overcome losses due to friction, and to allow for the imperfect efficiency of our motors and generators that convert electrical energy to gravitational energy and back.

The device we describe has been given various names. Arthur Clarke, in *The Fountains of Paradise* (Clarke, 1979), termed it a space elevator. I, in *The Web Between the Worlds* (Sheffield, 1979), called it a beanstalk. Other names include skyhook, heavenly funicular, anchored satellite, and orbital tower.

The basic idea is very simple. There are, however, some interesting "engineering details."

First, a cable can't simply run down from a position at geosynchronous height. Its own mass, acted on by gravity, would pull it down to Earth. Thus there must be a compensating mass out beyond geosynchronous orbit. That's easy enough; it can be another length of cable, or if we prefer it a massive ballast weight such as a captured asteroid.

Second, if we string a cable from geostationary orbit to Earth it makes no sense for it to be of uniform cross section. The cable needs to support only the length of itself that lies below it at any height. Thus the cable should be thickest at geosynchronous height, and taper to thinner cross sections all the way down to the ground.

What shape should the tapering cable be? In practice, any useful cable will have to be strong enough to stand the added weight of the payload and the lift system, but let us first determine the shape of a cable that supports no more than its own weight. This is a problem in static forces, with the solution (skip the next half page if you are allergic to equations):

$$A(r) = A(R) \cdot \exp \ (K \cdot f(r/R) \cdot d/TR)$$

In this equation, $A(r)$ is the area of the cable at distance r from the center of the Earth, $A(R)$ is the area at distance R of geosynchronous orbit, K is the Earth's gravitational constant, d is the density of cable material, T is the cable's tensile strength, and f is the function defined by:

$$f(x) = (3/2 - 1/x - x^2/2)$$

The form of the equation for $A(r)$ is crucial. First, note that the *taper factor* of the cable, which we define as $A(r)/A(R)$, depends only on the ratio of cable tensile strength to cable density, T/d, rather than actual tensile strength or density. Thus we should make a beanstalk from materials that are not only strong, but light. Moreover, the taper factor depends exponentially on T/d. If a cable originally had a taper factor from geosynchronous orbit to Earth of 100, and if we could somehow double the strength-to-density ratio, the taper factor would be reduced to 10. If we could double the strength-to-density again, the taper factor would go down to 3.162 (the square root of 10). Thus the strength-to-density ratio of the material used for the cable is enormously important. We note here the presence of the exponential form in this situation, just as we observed it in the problem of rocket propulsion.

We have glossed over an important point. Certainly, we know the shape of the cable. But is there any material with a large enough strength-to-density ratio? After all, at an absolute minimum, the cable has to support 35,770 kilometers of itself. The problem is not quite as bad as it sounds, since the Earth's gravitational field diminishes as we go higher. If we define the "support length" of a material as the length of uniform cross section able to be supported in a one-gee gravitational field, it turns out that the support length needed for the beanstalk cable is 4,940 kilometers. Since the actual cable can and should be tapered, a support length of 4,940 kilometers will be a good deal more than we need. On the other hand, we must hang a transportation system onto the central cable, so there has to be more strength than required for the cable alone.

Is there anything strong enough to be used as a cable for a beanstalk? The support lengths of various materials are given in TABLE 8.1 (p. 227).

The conclusion is obvious: today, no material is strong enough to form the cable of a beanstalk from geostationary orbit to the surface of the Earth.

However, we are interested in science fiction, and the absolute limits of what might be possible. Let us recall Chapter 5, and the factors that determine the limits to material strength. Examining TABLE 5.1 (p. 122), we see that solid hydrogen would do nicely for a beanstalk cable. The support length is about twice what we need. It would have a taper factor of 1.6 from geosynchronous orbit to Earth. A cable one centimeter across at the lower end would mass 30,000 tons and be able to lift payloads of 1,600 tons to orbit.

Unfortunately, solid metallic hydrogen is not yet available as a construction material. It has been made as a dense crystalline solid at room temperature, but at half a million atmospheres pressure. We need to have faith in progress. There are materials available, today, with support lengths ten times that of anything available a century ago.

Beanstalks are easier for some other planets. TABLE 8.2 (p. 228) shows what they look like around the solar system, assuming the hydrogen cable as our construction material.

Mars is especially nice. The altitude of a stationary orbit is only half that of the Earth. We can make a beanstalk there from currently available materials. The support length is 973 kilometers, and graphite whiskers comfortably exceed that.

Naturally, the load-bearing cable is not the whole story. It is no more than the central element of a beanstalk that will carry materials to and from orbit. The rest of the system consists of a linear synchronous motor attached to the load-bearing cable. It will drive payloads up and down. Some of the power expended lifting a load is recovered when we lower a similar load back down to Earth. The fraction depends on the efficiency of conversion from mechanical to electrical energy.

So far we have said nothing about actual construction methods. It is best to build a beanstalk from the top down. An abundant supply of suitable materials (perhaps a relocated carbonaceous asteroid) is placed in geostationary orbit. The load-bearing cable is formed and simultaneously extruded upward and downward, so that the total up and down forces are in balance. Anything higher than geosynchronous altitude exerts a net outward force, everything below geosynchronous orbit exerts a net inward force. All forces are tensions, rather than compressions. This is in contrast to what we may term the "Tower of Babel" approach, in which we build up from the surface of Earth and all the forces are compressions.

After extruding 35,770 kilometers of cable downward from geostationary orbit, and considerably more upward, the lower end at last reaches the Earth's equator. There it is tethered, and the drive train added. The beanstalk is ready for use as a method for taking payloads to geosynchronous orbit and beyond. A journey from the surface to geosynchronous height, at the relatively modest speed of 300 kilometers an hour, will take five days. That is a lot slower than a rocket, but the trip should be far more restful.

The system has another use. If a mass is sent all the way out to the end of the cable and then released, it will fly away from Earth. An object released from 100,000 kilometers out has enough speed to be thrown to any part of the solar system. The energy for this, incidentally, is free. It comes from the Earth itself. We do not have to worry about the possible effects of that energy depletion. The total rotational energy of the Earth is only one-thousandth of the planet's gravitational self-energy, but that is still an incredibly big number.

The converse problem needs to be considered: What about the effects of the Earth on the beanstalk?

Earthquakes sound nasty. However, if the beanstalk is tethered by a mass that forms part of its own lower end, the situation will be stable as long as the force at that point remains "down." This will be true unless something were to blow the whole Earth apart, in which case we might expect to have other things to worry about.

Weather will be no problem. The beanstalk presents so small a cross-sectional area compared with its strength that no imaginable storm can trouble it. The same is true for perturbations from the gravity of the Sun and Moon. Proper design will avoid any resonance effects, in which forces on the structure might coincide with natural forcing frequencies.

In fact, by far the biggest danger that we can conceive of is a man-made one: sabotage. A bomb exploding halfway up a beanstalk would create unimaginable havoc in both the upper and lower sections of the structure.

The descent of a shattered beanstalk was described, in spectacular fashion, in Kim Stanley Robinson's *Red Mars* (Robinson, 1993). My only objection is that in the process the town of Sheffield, at the base of the beanstalk, was destroyed.

8.22 Theme and Variations. We now offer three variations on the basic beanstalk theme. None needs any form of propellant or uses any form of rocket, and all could, in principle, be built today.

The *rotating beanstalk* is the brainchild of John McCarthy and Hans Moravec, both at the time at Stanford University. Moravec and McCarthy termed the device a *nonsynchronous skyhook*, though I prefer rotating beanstalk. It is a strong cable, 8,500 kilometers long in one design, that rotates about its center of mass as the latter goes around the Earth in an orbit 4,250 kilometers above the surface. Each end dips into the atmosphere and back out about once an hour.

The easiest way to visualize this rotating structure is to imagine that it is one spoke of a great wheel that rolls around the Earth's equator. The end of the beanstalk touches down like the spoke of a wheel, vertically, with no movement relative to the ground. Payloads are attached to the end of the beanstalk at the moment when it touches the ground. However, you have to be quick. The end comes in at about 1.4 gees, then is up and away again at the same acceleration.

The great advantage of the rotating beanstalk is that it can be made with materials less strong than those needed for the "static" beanstalk. In fact, it would be possible to build one today with a taper factor of 12, using graphite whiskers in the main cable. There is of course no need for such a structure to be in orbit around the Earth. It could sit far out in space, providing a method to catch and launch spacecraft.

The *dynamic beanstalk* has also been called a *space fountain* and an *Indian rope trick*. It is another elegant use of momentum transfer.

Consider a continuous stream of objects (say, steel bullets) launched up the center of an evacuated vertical tube. The bullets are fired off faster than Earth's escape velocity, using an electromagnetic accelerator on the ground. As the bullets ascend, they will be slowed naturally by gravity. However, they will receive an additional deceleration through electromagnetic coupling with coils placed in the walls of the tube. As this happens, the bullets transfer momentum upward to the coils. This continues all the way up the tube.

At the top, which may be at any altitude, the bullets are slowed and brought to a halt by electromagnetic coupling. Then they are reversed in direction and allowed to drop down another parallel evacuated tube. As they fall they are accelerated downward by coils surrounding the tube. This again results in an upward transfer of momentum from bullets to coils.

At the bottom the bullets are slowed, caught, given a large upward velocity, and placed back in the original tube to be fired up again. We thus have a continuous stream of bullets, ascending and descending in a closed loop.

If we arrange the initial velocity and the bullets' rate of slowing correctly, the upward force at any height can be made to match the total downward gravitational force of tube, coils, and anything else we attach to them. The whole structure will stand in dynamic equilibrium, and we have no need for any super-strong materials.

The dynamic beanstalk can be made to any length, although there are advantages to extending it to geosynchronous height. Payloads raised to that point can be left in orbit without requiring any additional boost. However, a prototype could stretch upward just a few hundred kilometers, or even a few hundred meters. Seen from the outside there is no indication as to what is holding up the structure, hence the "Indian rope trick" label.

Note, however, that the word "dynamic" must be in the description, since this type of beanstalk calls for a continuous stream of bullets, with no time out for repair or maintenance. This is in contrast to our static or rotating beanstalks, which can stand on their own without the need for continuously operating drive elements.

8.23 The launch loop. As we have described the dynamic beanstalk, the main portions are vertical, with turnaround points at top and bottom. However, when the main portion is horizontal we have a launch loop.

Imagine a closed loop of evacuated tube through which runs a continuous, rapidly moving metal ribbon. The tube has one section that runs from west to east and is inclined at about 20 degrees to the horizontal. This leads to a 2,000-kilometer central section, 80 kilometers above the Earth's surface and also running west to east. A descending west-to-east third section leads back to the ground, and the fourth section is one at sea level that goes east to west and returns to meet the tube at the lower end of the first section.

The metal ribbon is 5 centimeters wide and only a couple of millimeters thick, but it travels at 12 kilometers a second. Since the orbital velocity at 80 kilometers height is only about 8 kilometers a second, the ribbon will experience a net outward force. This outward force supports the whole structure: ribbon, containing tube, and an electromagnetic launch system along the 2,000 kilometer upper portion of the loop. This upper part is the acceleration section, from which 5-ton payloads are launched into orbit. The whole structure requires about a gigawatt of power to maintain it. Hanging cables from the acceleration section balance the lateral forces produced by the acceleration of the payloads.

Although the launch loop and the dynamic beanstalk both employ materials moving through evacuated tubes, they differ in important ways. In the dynamic beanstalk the upward transfer of momentum is obtained using a decelerating and accelerating particle stream. By contrast, the launch loop contains a single loop of ribbon moving at constant speed and the upper section is maintained in position as a result of centrifugal forces.

8.24 Space colonies. I can imagine some readers at this point saying, all this talk of going to space and traveling in space, and no mention of space colonies except those on the surface of planets. There are hundreds and hundreds of stories about self-sufficient colonies in space.

There are indeed, and during the 1970s I read many of them with pleasure and even wrote some myself. One of the most fruitful ideas involved "L-5 colonies." "L-5" describes not a type of colony, but a *place*. In the late eighteenth century, the great French mathematician Joseph Louis Lagrange studied the problem of three bodies orbiting about each other. This is a special case of the general problem of N orbiting bodies, and as mentioned in the previous chapter, no exact solution is known for N greater than 2. Lagrange could not solve the general 3-body problem, but he could obtain useful results in a certain case, in which one of the bodies is very small and light compared with the other two. He found that there are five places where the third body could be placed, and the gravitational and centrifugal forces on it would exactly cancel. Three of those places, known as L-1, L-2, and L-3, lie on the line joining the centers of the two larger bodies. The other two, L-4 and L-5, are at the two points forming equilateral triangles with respect to the two large bodies, and lying in the plane defined by their motion about each other.

The L-1, L-2, and L-3 locations are unstable. Place a colony there, and it will tend to drift away. However, the L-4 and L-5 locations are stable. Place an object there, and it will remain. There are planetoids, known as the Trojan group, that sit in the L-4 and L-5 positions relative to Jupiter and the Sun.

The Earth-Moon system also has Lagrange points, which in the case of the L-4 and L-5 points are equidistant from Earth and Moon. In the 1970s, an inventive and charismatic Princeton physicist, Gerard O'Neill, proposed the L-5 location as an excellent place to put a space colony (L-4 would actually do just as well). The colonies that he designed were large rotating cylinders, effective gravity being provided by the centrifugal force of their rotation. Within the cylinder O'Neill imagined a complete and self-contained world, with its own water, air, soil, and plant and animal life. Supplies from Earth or Moon would be needed only rarely, to replace inevitable losses due to small leaks.

The idea was a huge success. In 1975 the L-5 Society

was formed, to promote the further study and eventual building of such a colony.

What has happened since, and why? Gerard O'Neill is dead, and much of his vision died with him. The L-5 Society no longer exists. It merged with the National Space Institute to become the National Space Society, which now sees its role as the general promotion of space science and space applications.

More important than either of these factors, however, is another one: economic justification. The prospect of a large self-sufficient space colony fades as soon as we ask who would pay for it, and why. Freeman Dyson (Dyson, 1979, Chapter 11) undertook an analysis of the cost of building O'Neill's "Island One" L-5 colony, comparing it with other pioneering efforts. He made his estimate not only in dollars, but in cost in man-years per family. He decided that the L-5 colony's per family cost would be hundreds of times greater than other successful efforts. He concluded "It must inevitably be a government project, with bureaucratic management, with national prestige at stake, and with occupational health and safety regulations rigidly enforced." All this was before the International Space Station, whose timid builders have proved Dyson exactly right: "The government can afford to waste money but it cannot afford to be responsible for a disaster."

The L-5 colony concept has appeal, and the technology to build the structure will surely become available. But it is hard to see any nation funding such an enterprise in the foreseeable future, and still harder to imagine that industrial groups would be interested.

The L-5 colony—regrettably, because it is such a neat idea—is part of what I like to call false futures of the past, projections made using past knowledge that are invalidated by present knowledge.

I believe there will certainly be space colonies in the future. Write stories about them by all means. But don't make them rotating cylinders at the L-5 location. Those stories have already been written.

8.25 Solar power satellites. While in skeptical mode,

let me say a few words about another concept of initial high appeal, the Solar Power Satellite. This was proposed in the 1960s by Peter Glaser, and like the L-5 colonies it had its heyday in the 1970s and early 1980s. Proponents of the idea believed (and believe) that it can help to solve Earth's energy problems.

A solar power satellite, usually written as SPS, has three main components. First, a large array of photoreceptors, kilometers across, in space. Each receptor captures sunlight and turns it to electricity. The most usual proposed location is in geosynchronous orbit, though some writers prefer the Earth-Moon L-4 location. The second component is a device that converts electricity to a beam of microwave radiation and directs it toward Earth. The third component is a large array on the surface of the Earth, usually known as a *rectenna,* that receives the microwave radiation and turns it into electricity for distribution nationally or internationally.

The SPS has some great virtues. It can be placed where the Sun is almost always visible, unlike a ground-based solar power collector. It taps a power source that will continue to be steadily available for billions of years. It contributes no pollution on Earth, nor does it generate the waste heat of other power production systems. It does not depend on the availability of fossil or nuclear fuels.

Of course, the SPS cannot be built without a powerful in-space manufacturing capability, something that is lacking today. We are having trouble putting modest structures, such as the International Space Station, into low orbit. It is likely that we will not be able to build an object as large as the proposed SPS for another century or more.

But when a century has passed, we are likely to have much better energy-raising methods, such as controlled fusion. Admittedly, progress on fusion has been slow—we have been promised it for fifty years—but it, or some other superior method, will surely come along. A fusion plant (or, for that matter, a fission plant) *in orbit* would have all the advantages of SPS, and none of the disadvantages. Sunlight is a highly diffuse energy source unless you get very close to the Sun. As we pointed out in Chapter 5, the history of energy use shows a move in the direction of more compact power sources—oil is more intense and compact than water or wind, nuclear is more compact and intense than chemical. The other problem is that the Sun, unlike our future fusion reactors, was not designed to fit in with human energy uses and needs. I put the question the other way round: Why build a kilometers-wide array, delicate and cumbersome and vulnerable to micrometeor damage, when you can put the same power generating capacity into something as small as a school bus? Admittedly, we don't have controlled fusion yet—but we also can't build an SPS yet.

However, the real killer argument is not technological, but economic. Suppose you launch SPS to serve, say, the continent of Africa. You still have the problem, who will pay for the energy? Economists distinguish two kinds of demand: real demand: the need for food of starving people with money to buy it; and other demand: the need for food of starving people without money. Regrettably, much demand for energy is in nations with no resources to pay for it.

In spite of this economic disconnect, many people have suggested that an SPS would be great for providing energy to Africa, where energy costs are high. Suppose that you put SPS is geostationary orbit and beam down, say, 5 gigawatts. That's the power delivered by a pretty substantial fossil fuel station. Now, you could also generate that much energy by building a dam on the Congo River, where it drops sharply from Kinshasa to the Atlantic. So ask yourself which you would prefer if you were an African. Would you like SPS, providing power from a source over which you had no control at all—you couldn't even get to visit it. Or would you prefer a dam, which in spite of all its defects, sits on African soil and is at least in some sense under your control? SPS has to compete not only from an economic point of view, but from a social and political point of view.

I think it fails on all those counts. Like the L-5 colony, SPS is part of a false future. It is not surprising to find Gerard O'Neill arguing that the sale of electricity generated by an SPS at L-5 would pay for the colony in the breathtakingly short period of twenty-four years. When we want to do something, all our assumptions are optimistic.

There are still SPS advocates. A recent NASA study suggested that a 400 megawatt SPS could be built and launched for five billion dollars. Do I believe that number? Not in this world. We all know that paper studies often diverge widely from reality. NASA's original estimated cost to build the International Space Station was eight billion dollars. Over the years, the station has shrunk in size and the costs have risen to more than 30 billion dollars. Projects look a lot easier before you get down to doing them. Recall the euphoria for nuclear power plants in the 1940s, "electricity too cheap to meter." And that was for something we had a lot more experience with than the construction of monster space structures.

Certainly, we hope and expect that the cost of sending material to space will go down drastically in the next few generations. We also will become increasingly unwilling to pollute the Earth with our power generation. But frequent space launches have their own effects on the environment of the upper atmosphere. If there is ever an SPS, which I doubt, it will more likely make little use of Earth materials and depend on the prior existence of a large space infrastructure.

I feel sure that will come—eventually. By that time the idea of power generation plants near population centers will be as unacceptable as the Middle Ages habit of allowing the privy to drain into the well. However, I want to emphasize that our solutions to the problems of the future can be expected to work no better than two-hundred-year-old solutions to the problems of today. We can propose for our distant descendants our primitive technology as fixes for their problems. But I don't believe that they will listen.

TABLE 8.1
Strength of materials.

Material	Density (gms/cc)	Tensile strength (kgms/sq.cm.)	Support length (kms)
Lead	11.4	200	0.18
Gold	19.3	1,400	0.73
Aluminum	2.7	2,000	7.40
Cast iron	7.8	3,500	4.50
Carbon steel	7.8	7,000	9.00
Manganese steel	7.8	16,000	21.00
Drawn steel wire	7.8	42,000	54.00
Kevlar	1.4	28,000	200.00
Iron whisker	7.8	126,000	161.00
Silicon whisker	3.2	210,000	660.00
Graphite whisker	2.0	210,000	1,050.00

TABLE 8.2
Beanstalks around the solar system.

Body	Radius of stationary satellite orbit (kms)	Taper factor
Mercury	239,731	1.09
Venus	1,540,746	1.72
Earth	42,145	1.64
Luna	88,412	1.03
Mars	20,435	1.10
Jupiter	159,058	842.00
Callisto	63,679	1.02
Saturn	109,166	5.11
Titan	72,540	1.03
Uranus	60,415	2.90
Neptune	2,222	6.24
Pluto*	20,024	1.01

* Since Pluto's satellite, Charon, seems to be in synchronous orbit, a beanstalk directly connecting the two bodies is feasible.

CHAPTER 9
Far-Out Alternatives

9.1 Problems of interstellar travel. One of the strongest of today's limitations on science fiction writers is the pesky constancy of the speed of light. If you can't go faster than that, light-speed limitation is—to put it mildly—an inconvenience for travel to even the nearest stars.

To many people, travel to the stars may not seem so difficult. After all (the logic goes) a dozen humans have already been to the Moon and back. We have sent landers to Mars, and we plan to do so again. Our unmanned probes have allowed us to take a close look at every planet of the Solar System except Pluto.

After interplanetary travel surely comes interstellar travel. If we have been able to do so much in the forty years since the world's space programs began, shouldn't an interstellar mission be possible in a reasonable time . . . say, thirty or forty years from now?

In a word, no.

For travel on Earth, different transportation systems can be nicely marked by factors of ten. Up to two miles, most of us are (or should be) willing to walk. For two to twenty miles, a bicycle is convenient and reasonable. A car is fine from twenty to two hundred, and above that most of us would rather fly or take a train.

Away from Earth, the factor of ten is no longer convenient. Our closest neighbor in space, the Moon, is about 240,000 miles away, or 400,000 kilometers. A factor of ten does not take us anywhere interesting. Nor does a factor of a hundred. We have to use a factor of 1,000 to take us as far as the Asteroid Belt.

Ten thousand times the distance to the Moon takes us four billion kilometers from Earth, to the outer planets of the Solar System. We are still a long way from the stars. For that we need another factor of 10,000. Forty trillion kilometers is about 4.2 light-years, and that is close to the distance of Alpha Centauri. Thus, the nearest star is about 100,000,000 times as far away as the Moon.

Want to visit the center of our galaxy, a common drop-in point for science fiction travelers? That is almost 10,000 times as far away as Alpha Centauri, a trillion times as far away as the Moon. Getting to the Moon, you may recall, was considered a big deal.

Numbers often have little direct meaning. Perhaps a more significant way of thinking of the distance to the stars is to imagine that we have a super-transportation system, one that can carry a spacecraft and its crew to the Moon in one minute. Anyone interested in solar system development would drool at the very thought of such a device. Yet we will take 190 years to reach Alpha Centauri, and most of the stars that we think of as "famous" are much farther away: 1,300 years trip time to Vega, over 20,000 years to Betelgeuse. Galactic center? Sorry, that's going to take a couple of million years.

We need a faster-than-light drive. But before we consider exotic alternatives, let's take one more look at what we might do within the confines of the laws of physics as they are known today. If nothing can travel faster than light, can we use light itself as our tool?

We can, provided we are willing to send and receive signals, rather than material objects. That is what SETI, the Search for Extraterrestrial Intelligence, is all about.

9.2 The Search for Extraterrestrial Intelligence. ET made it look easy. You collect a few bits and pieces of electronics, join them together in some mysterious way, and lo and behold, you have a transmitter that will send a signal to the stars. You switch on, and wait for your friends to show up.

ET did not ask the assistance of Earth scientists in sending his message; but suppose that he had. Suppose that we were asked to send a message to the stars, one that could be received and interpreted many light-years away. What techniques would we choose, and how would we go about it?

The idea of sending messages to beings on other worlds is an old one. In the 1820s, the mathematician Carl Friedrich Gauss proposed to lay out huge geometrical figures on the surface of Earth. He argued that these, seen through telescopes by the inhabitants of other planets, would give proof that Earth harbored intelligent life. The principal pattern, created by the layout of large fields containing crops of different colors, would show a right-angled triangle with each side bordered by squares. We would provide graphic evidence that Earthlings (though not, apparently, many of America's high school students) are familiar with the theorem of Pythagoras.

Gauss had in mind the nearer planets of the solar system, since even with big telescopes the biggest fields on Earth could not be seen from farther away than Mars or Venus. Nonetheless, given its limitations, Gauss's idea is not impossible. It represents wonderfully advanced thinking for its time.

Similar suggestions involving the lighting of great fires in the Sahara Desert were made later in the nineteenth century. By 1900, extraterrestrial communication had become a popular subject. In that year the French Academy of Sciences offered a prize of 100,000 francs to the first person making contact with another world. The planet Mars was specifically excluded, since that was considered too easy.

These early proposals for extraterrestrial communications all had one thing in common: they assumed that visible light would be the best way to communicate over great distances. At first that seems a fair assumption, even when we extend our goal from interplanetary to interstellar space talk. We live on a planet orbiting a fairly typical star. Our eyes have evolved to be sensitive to the light of that star, as modified by passage through the Earth's atmosphere. Other beings, born on planets that circle other stars, are likely to have developed organs of sight. It would be most efficient for them to have developed maximum sensitivity in roughly the same wavelength region as us. Therefore we should be able to communicate by optical techniques, using the part of the electromagnetic spectrum visible to humans.

This sounds reasonable, but it misses a key point. Visible wavelengths are not the best ones for interstellar communication, precisely because visible light is so abundant throughout the universe. We can certainly send a signal, but another being will have trouble distinguishing it from natural signals that every planet, star, and galaxy emits or reflects at the same wavelengths.

Detection would be a formidable task. There is just too much clutter in the spectral window between 0.40 and 0.70 micrometers, where we ourselves see. Our message will be lost in the background noise that Nature is generating all around us.

What we need is a signal that will not be confused by emissions from stars, planets, interstellar dust clouds, galaxies, or any other natural source in the universe. We must find a "quiet" part of the spectrum, in which Nature does not make strong signals of her own; and we need a region where other beings would find logical reasons to send and look for signals.

This sounds like a difficult proposition, but fortunately such a region does exist.

9.3 The choice of signal carrier. If we sit down to make a list of the properties that any signalling system should have for communication over interstellar distances, we find that our signal must satisfy these requirements:

1) It should possess characteristics that allow it to be readily distinguished from naturally generated emissions;

2) It should not be easily absorbed by interstellar dust and gas;

3) It should be easy to detect;

4) It should be easy to generate with modest amounts of power;

5) It should travel at high speed.

We assume that no signal can travel faster than the speed of light, so anything traveling at light-speed will be our first preference.

That at once rules out certain signaling methods. For example, the Pioneer 10 and 11 and the Voyager 1 and 2 spacecraft are on trajectories carrying them out of the solar system. They are on their way to the stars, and they even contain messages intended for other beings. However, they travel horribly slowly. It will be hundreds of thousands of years before they reach the nearest stars. Thus they, and any other spacecraft described in Chapter 8, are too slow for interstellar messages.

The speed requirement does nothing to limit our choice within the electromagnetic spectrum. Everything from X-rays and gamma rays to visible light and long-wavelength radio waves travel in vacuum at the same speed; our other four criteria must be employed to select a preferred wavelength.

The first systematic examination of the whole spectrum, to see what is best for interstellar communication, was done by Philip Morrison and Giuseppe Cocconi (Morrison and Cocconi, 1959). As Morrison has remarked, they started out thinking that gamma rays would be the best choice, and only later broadened their viewpoint to include the whole electromagnetic spectrum.

After making their study, they concluded that there are indeed preferred wavelengths for interstellar communication, wavelengths that in fact satisfy all five of the criteria listed above. Morrison and Cocconi also addressed the question of how the signals might be generated and received.

Left out of consideration—deliberately—was the question of who might be sending signals to us. As Morrison put it, informally, "See, you were thinking that in order to call somebody up, you have to have somebody to call. I'm saying that before you call, you have to have a telephone system. We got our initial idea from the telephone system, not from thinking that anyone is there. We don't know how to estimate the probability of extraterrestrial intelligence . . . but if we never try, we'll never find it."

The inability to estimate that probability has not stopped people from trying. Suppose we write an equation giving the number of technologically advanced civilizations sending out messages in our galaxy as a product of seven independent factors: 1) the number of stars in our galaxy; 2) the fraction of such stars with planets; 3) the average number of planets orbiting any star that are suitable for the development of life; 4) the fraction of planets where life actually develops; 5) the fraction of life-bearing planets that develop intelligent life; 6) the fraction of intelligent life forms who actually seek to communicate with other forms; and 7) the fraction of the planet's lifetime occupied in the communicating phase.

This is known as the Drake Equation. It was proposed by Frank Drake, often considered the father of SETI. In 1960, using an 85-foot radio telescope in Green Bank, West Virginia, he was the first person to seek radio signals from extraterrestrial intelligences.

There are a few things to note about this equation. First, it is not a physical law, but merely an enumeration of factors. Second, if any factor is zero, the left hand side and hence the number of signals is zero. Third, only the first factor, the number of stars in our galaxy, is known to even one significant figure. The rest are little more than blind guesses.

Although thousands of pages have been written about the Drake Equation and its factors, I don't think it tells us much. The right attitude was expressed by Freeman Dyson, in his book *Disturbing the Universe* (Dyson, 1979): "I reject as worthless all attempts to calculate from theoretical principles the frequency of occurrence of intelligent life forms in the universe . . . Nevertheless, there are good scientific reasons to pursue the search for evidence of intelligence. . . ."

Morrison and Cocconi examined the whole electro-magnetic spectrum. They reported their results in *Nature* magazine, and asserted that the microwave region, the one that we use for terrestrial radio and radar, is the best place to put your signal. This wavelength regime is markedly quieter (less cluttered by natural signals) than the gamma ray, X-ray, ultraviolet, visible, or infrared ranges. Nature seems to have overlooked this region for stars and planets, to the point where Earth, with its copious emissions of man-made radar, radio, and television signals, is by far the most powerful source in the solar system. At microwave wavelengths Earth is brighter than Jupiter or even the sun, although the latter is a beacon millions of times brighter at visible wavelengths.

Further, even within the microwave region, there is a definite preferred window, a "quiet spot" between 30 centimeter wavelength (1 gigaHertz frequency) and 0.3 centimeter wavelength (100 gigaHertz frequency). Wavelength and frequency are inversely related, since frequency times wavelength=the speed of light. Thus either wavelength or frequency can be used equally well to define a range of the spectrum. When we speak of radio or radar we usually work in terms of frequencies; for visible or infrared light, we generally use wavelengths.

If we want to send or receive signals from the surface of the Earth, rather than out in space, then the absorption properties of our atmosphere must be taken into account. We also have to note that man-made signals from radio and television and radar form a possible source of noise for external signals. This finally reduces the quietest region to a "terrestrial microwave window" from 1 to 10 gigaHertz (30 to 3 centimeters).

Below 1 gigaHertz, the natural synchrotron radiation of the galaxy provides unwanted noise. Above 20 gigaHertz, the quantum noise of spontaneous emission dominates; but between 1 and 10 gigaHertz the only significant noise is the cosmic background radiation, peaking at a temperature of 2.7 Kelvin and an associated frequency of 25 gigaHertz, but still appreciable between 1 and 10 gigaHertz.

By fortunate coincidence, conveniently within this valley of quiet lie two significant spectral lines: at 1.420 gigaHertz (21 centimeters) we find the radiation emission of neutral hydrogen, and at 1.662 gigaHertz (18 centimeters) the emission of the hydroxyl radical. Together, hydrogen and the hydroxyl radical combine to form water, the basis for all life as we know it. As Project Cyclops, an early study of search methods for extraterrestrial intelligence, stated with memorable imagery:

"Nature has provided us with a rather narrow band in this best part of the spectrum that seems specially marked for interstellar contact. It lies between the spectral lines of hydrogen (1420 megaHertz) and the hydroxyl radical (1662 megaHertz). Standing like the Om and the Um on either side of a gate, these two emissions of the disassociation products of water beckon all water-based life to search for its kind at the age-old meeting place of all species: the water hole" (1972; cited in NASA SP-419, 1977, edited by Philip Morrison).

If we are going to use radiation to send our interstellar signal then this place, the "water hole," provides the best set of frequencies. Moreover, signals in this region can be generated easily, with standard radio equipment; they can be beamed in any direction that we choose; and they will be detectable over stellar distances with the transmission power available to us today.

There is still a problem: deciphering a possible message. A signal is not acceptable as artificial (remember the pulsars) until it is decoded. Of course, a "message" in the usual sense is not needed; it would be quite sufficient if the pulses that we receive were, say, the prime numbers, or numbers followed by their squares.

In the early days of SETI, Frank Drake devised a short message containing some basic information about us. He sent it to a number of his colleagues, telling them that it *was* a message and inviting them to decipher it. Not one of them succeeded. Can you? Drake's "message" is given in TABLE 9.1 (p. 246).

The messages sent out on the Voyager spacecraft had the same problem. They included music, the sound of rain and cars, and a statement from President Jimmy Carter; the sign of intelligence, perhaps, but one difficult to interpret, even for its senders.

Note the difference between detecting extraterrestrial signals, and sending signals for others to receive. These two different problems are often confused, but SETI is the search for extraterrestrial intelligence (we sit and listen, but we don't send any signals ourselves), and CETI is communication with extraterrestrial intelligence (we also send our own messages).

The same instruments may be used either to send or to receive signals. A radio telescope can listen, by placing detection equipment at its focal point; or it can send, by placing a transmitter at the same focal point. The signal can be sent to any preferred direction in space.

The 1,000-foot radio telescope at Arecibo in Puerto Rico has been used in both modes; to listen for signals from many places, and to beam a coded signal to the Hercules globular star cluster, M13, 25,000 light-years from Earth. A radio telescope a little bigger than the one at Arecibo Observatory would be able to detect that same signal when it reaches M13, 25,000 years from now.

The big problem with SETI was stated with admirable succinctness by the great Italian scientist, Enrico Fermi: *Where are they?* If there are extraterrestrial intelligences, some are presumably more advanced than we are. Why haven't they showed up and presented themselves at the United Nations, or sent a proof of their existence that is impossible to miss or deny?

This absence of contact is known as the *Fermi Paradox*, though it is hardly a paradox. It is simply a good question, to which there is no good answer. Some people suggest that we have not yet found the right radio frequency, or have looked in the wrong direction. Some argue that we are still in too primitive a state of technology, so that our proposed methods of sending or receiving signals are little better than the multicrop agricultural fields proposed by Gauss. And of course there are others who say that aliens don't need to signal, because they visit Earth on a daily basis in UFOs.

But then there is the alternative viewpoint: We are alone, the only intelligent species in our galaxy. It is a waste of time and money scanning the sky for messages, or sending them out to nowhere.

Is SETI a waste of time, as its critics say, because the probability of success is low? Or is it, as its disciples claim, a project that we ought to be engaged in all the time and at an increased level of effort, because the payoff of success could be so enormous?

Although the United States Senate cut off all SETI funding in 1993, the effort continues with private support. The program that used to be at NASA Ames has moved, almost in its entirety, to the SETI Institute in Mountain View, CA. The Planetary Society, in Pasadena, continues an active search under Paul Horowitz at Harvard. A very readable background discussion of the whole subject can be found in *The Search for Extraterrestrial Intelligence: Listening for Life in the Cosmos* (McDonough, 1987).

In science fiction, SETI has long been an accepted element of the field. Three good and very different examples are *The Hercules Text* (McDevitt, 1986); *Contact* (Sagan, 1985); and *The Ophiuchi Hotline* (Varley, 1977).

9.4 Beam me up. We know how to send signals to other civilizations at the speed of light, and we are already looking for messages from them. But do we really need to go in person? Why not transmit a complete signal that represents you or me, and use it to re-create us at the other end? That way, we'll get to the stars as fast as possible, and so far as subjective experience is concerned it will be no time at all.

Put aside for one moment the fact that there is nothing at the other end to put us back together. Ignore also the awkward question of which one is the real you— the one who was scanned back here, or the one who is reconstructed out there. Let us size the problem.

The human body contains about 10^{28} atoms. To specify the substance of each atom (i.e. the element) needs only two decimal digits, since all atoms of a particular element are identical. However, we also need to specify information where each atom is. That calls for three coordinates, each given to an accuracy of, say, 2×10^{-10} meters, and with a maximum value of a couple of meters (basketball players have to crouch, or stay home).

Associated with each coordinate we specify a number from 1 to 99 (for the appropriate element) with a zero when a coordinate lies outside your body. A representation of a complete human, down to hair part, birthmarks, and eye color, thus calls for about 10^{32} separate pieces of information. We assume that we have an understanding, in advance, that the (x,y,z) coordinates of atoms will be given sequentially, in a particular order.

Let's see how long it will take to transmit a person from one place to another (distance is not relevant). A high-speed data link from ground to space is a few hundred million bits per second. We will be generous, and say we have a data link of a billion decimal digits (10^9) per second. Then the transfer of one human will take 10^{23} seconds, or 10^{15} years. The universe is only about 10^{10} years old. We would be better off using the Post Office.

All right, we will seek economies. First, is it really necessary to send an exact description of every atom? As we know from heart, kidney, and liver transplants, these organs are all functionally similar. Let us send just the information defining the "real you," the brain and maybe a few glands that seem to define our emotions. We do not save very much. The range of each coordinate reduces from 2 meters to 20 centimeters. The transmission time comes down to about 10^{12} years. Still no good.

How about if we simply regard the brain as a computer, and download the information held in it? This is certainly a popular science fiction device, although Penrose, as we will discuss in some detail in Chapter 13, would argue that it is impossible because the brain is more than a computer.

Let us assume that he is wrong. There are about 10^{11} neurons in the brain. We number them sequentially, 1, 2, 3 . . . 10^{11}. We make the (disputed) assumption that a neuron is a simple on-off device, so that its information can be represented by a single binary digit. Further, we assume that each neuron connects to an average of 50 other neurons (this is only an average; certain neurons in the cerebellum have up to 80,000 connections). Now we assume that the brain is completely defined by the neuron contents, plus all the neuron-to-neuron connections. For every neuron, we need to specify a binary digit, plus 50 decimal numbers each of which may be up to 11 digits long. Then a human, regarded solely as information, is defined totally by 10^{14} decimal digits. The transmission time using our billion-digit-a-second transmission system is a little over a day.

This is an acceptable period. Even if we send the signal with triple redundancy, to make sure that the you-that-arrives is not subtly different from the you-that-was-sent, we are talking transmission times of a few days.

Note that you will not exist physically until you have been downloaded from signal form into a clone of your body. That will be grown from your unique DNA description, which requires only about ten billion binary digits (a small fraction of the total signal) and could be sent as a lead file to the main message.

What will we do when it turns out that the SETI signal is not the *Encyclopedia Galactica* at all, but the exact prescription for some alien interstellar tourist?

9.5 A helping hand from relativity. It is Einstein's special theory of relativity that tells us we can never accelerate any object to move faster than the speed of light. The same theory, curiously enough, offers a helping hand when we want to travel long distances.

As we mentioned in Chapter 2, one standard and experimentally tested consequence of relativity is *time dilation*. Let us recap its effects. When an object (in our case, a spacecraft) moves at close to the speed of light, time as measured onboard the spacecraft feels the same to the passengers, but as far as an external observer is concerned, it is slowed.

The rule is very simple: for an object traveling at a fraction F of the speed of light, when an interval T passes in the rest of the universe, an interval only $\sqrt{(1-F^2)}$ of T passes in the object's reference frame.

Thus, if a ship travels at 90 percent of light-speed, time onboard relative to the outside universe is slowed by a factor 0.43; when a century passes on Earth, only 43 years pass on the ship. At 99 percent of light-speed, 14 years pass on the ship; at 99.9 percent of light-speed, 4 years pass on board. Clearly, if we can accelerate the ship close enough to light-speed—no mean feat, as we have seen already—then so far as the passengers are concerned, travel time to the stars or even to remote parts of the galaxy can be made tolerable. As an example, the center of the galaxy is about 30,000 light-years away. For a ship that traveled just one hundred meters a second slower than the speed of light, the perceived travel time from here to the galactic center would be only 24 years.

Frank Tipler, grandly dismissing the practical details of ship drive design, has examined travel times for a ship that moves not at constant speed, but at constant acceleration (Tipler, 1996). Setting that acceleration at a comfortable one gee, Tipler finds that a round trip to the center of our galaxy will take about 40 years of shipboard time. The Andromeda Galaxy is about 2.2 million light-years away. A visit to it needs 57 years of ship time. And if we want to take a longer trip, to the Virgo Cluster at 60,000,000 light-years distance, we can expect to be away for about 70 years. As we see, a constant acceleration telescopes almost all distances down to the point where a human lifetime is enough to travel them.

The snag, of course, is that you might go to the center of the galaxy and back in one lifetime; but while you were gone, things here on Earth could be expected to change considerably in your 60,000-year absence. As for a trip to the Virgo Cluster, you could have left Earth when dinosaurs were the dominant land animals, and not be back yet.

9.6 Faster than light. Like it or not, we have to explore the possibilities of faster-than-light travel. Without it, all our interstellar empires and intergalactic trade shows are impossible. What's the point of sending your army to quell an uprising when it happened 50 centuries ago, or ordering a piece of furniture that will take a thousand years to be delivered?

We need a loophole. One possibility was suggested in Chapter 2, where the idea of quantum teleportation was explored. To find another one, let us return to Einstein. The assumption that we cannot travel faster than light is usually stated as one of the central elements of the theory of special relativity. In fact, what Einstein said was not quite that. You cannot *accelerate* an object faster than light, or even as fast as light. As you try to move something faster and faster, the energy needed to do it becomes greater and greater.

However, this does not mean that particles which travel faster than light cannot exist. As one researcher into faster-than-light particles pointed out, that would be like saying that there can be no people north of the Himalayas, since no one can climb over the mountain ranges.

In this case, the mountain range is the speed of light. Although no particle can be accelerated through the barrier, this in no way proves that particles cannot exist on the other side of the barrier.

In 1967, Gerald Feinberg gave a name to hypothetical faster-than-light particles. He called them *tachyons*, from the Greek word *tachys*, meaning swift. Richard Tolman, as early as 1917, thought he had proved that the existence of tachyons would allow information transfer to the past, and thus allow history to be changed. For example, a message back to 1963 could in principle have prevented the Kennedy assassination. That possible use of tachyons was explored in the novel *Timescape* (Benford, 1979). Today, however, Tolman's argument is no longer accepted; Benford's novel remains as fiction.

Tachyons do appear to be permissible within the framework of conventional physics, in that there seems to be no physical or logical law ruling out the possibility of their existence. This has led some writers to argue that tachyons must exist, adopting the rule of the anthill from *The Once and Future King* (White, 1958): "Everything not forbidden is compulsory."

Suppose for the moment that tachyons are real. Then to see how light speed forms a natural barrier separating bradyons (the familiar "slow" particles of our universe, also known as tardyons) from tachyons, imagine that we accelerate a charged particle faster and faster, for example using an electromagnetic field.

What happens to it? The particle certainly continues to increase in speed, but according to the theory of special relativity as it gets closer to the speed of light it also becomes more massive. As a result it becomes more difficult to accelerate. The mass doubles, then quadruples, and more and more energy is needed to speed it up just a little more.

The process never ends. To accelerate it to the speed of light would take an infinite amount of energy, and is therefore impossible.

In the same way, for a tachyon it takes more and more energy to slow it from above light speed. It would take an infinite amount of energy to slow it to the speed of light. Thus if both bradyons and tachyons exist, each is confined to its own velocity region. The speed of light is a "barrier" that forever separates the world of tachyons from the world of bradyons, and one can never become the other.

This is all logically self-consistent, but a big question remains: How could one detect the presence of a tachyon? It used to be thought that any charged particle traveling faster than light would emit a particular radiation, known as *Cherenkov radiation*; but this is no longer believed to be the case. The simplest way to detect a tachyon's presence is through a time-of-flight test: if two particle detectors each register an event, and the distance between them is so large and the times so close that only a particle travelling faster than light could cause them both, then we have a candidate tachyon.

This is a suspect mechanism for detection. If two people in a household come down with influenza within an hour of each other, we do not conclude that the incubation time for influenza must be one hour or less. Almost certainly, they both caught the flu from a third party. How would we ever know that a similar underlying cause did not lead to a false inference of tachyon presence?

9.7 Wormholes and loopholes. Let us assume that tachyons exist. We then have a possible way of sending messages faster than light. That's useful, but it's not enough. We want to send people between the stars, fast enough to offer the writer some storytelling freedom. Tachyons won't do that, they are signals only. They also have an unfortunate property of allowing those signals to travel backward in time, which might alone be enough reason to avoid them.

Where do we turn for plausible physics, a loophole that will allow us go faster than light without using tachyons or the still-unexplored world of quantum teleportation?

As is so often the case, we turn to the work of Albert Einstein. Thanks to the general theory of relativity, the structure of the universe is not a single, simply-connected region of space and time. As we saw in Chapter 3, no information from inside a black hole can ever reach the rest of the universe. This is still true, even when we allow for the Hawking evaporation process. Thus a black hole provides, in a very real sense, an edge of the universe. If black holes are common, then the whole universe has a curious kind of Swiss-cheese structure of holes and real (i.e. accessible) space.

Furthermore, there are regions close to a rotating black hole where very strange things can happen, at least in theory. Spacetime near the ring singularity of a kernel seems to be multiply connected. In other words, if you go close enough to the singularity, you may suddenly find yourself elsewhere, having been transported through a kind of spacetime tunnel. One problem is that you are likely to appear not only elsewhere, but elsewhen. The transport mechanism may also serve as a time machine.

Other forms of spacetime tunnels, known popularly as "wormholes," have been developed by Kip Thorne and fellow-workers at CalTech. We say "developed," but of course the development so far is purely conceptual. It calls for extracting a minute black hole from the enormous numbers continuously appearing and disappearing at distances of the Planck length (again see Chapter 2). The trick is then to stabilize one—it will try desperately to disappear—and inflate it to a size useful for transmission of human-sized objects. This calls for materials far stronger than anything we have today, although our positronium-positronium bonds of Chapter 5 are taking us in the right direction. The magnified, stabilized wormhole can then serve, like our kernel ring singularity, as a way to travel between distant points without traversing intermediate "normal" space. Again, there seems to be a substantial danger that the traveler will appear in elsewhen.

TABLE 9.1
Frank Drake's proposed message to the stars.

```
111100001010010000110010000000010000010100
100000110010110011110000011000011010000000
001000001000010000100010101010001000000000
000000000001000100000000000101100000000000
000000010001110110101101010000000000000000
000010010000111010101010000000000101010101
000000000111010101011101011000000001000000
000000000001000000000000001000100111111000
001110100000101100000111000000001000000000
100000000100000001111100000001011000101110
100000001100101111101011111000100111111001
000000000001111100000010110001111111100000
100000110000011000010000110000000011000101
001000111100101111
```

Got it? No, neither could I.

CHAPTER 10
Deus Ex Machina:
Computers, Robots,
Nanotechnology, Artificial Life,
and Assisted Thought

10.1 Computer limits. This chapter of the book is a hard one to write. The reason is simple: if we are given a large set of data points on a graph and asked to say where the next points are likely to lie, we can do a fairly good job. We fit curves to the given points, and extrapolate. If we have just a couple of points, however, the task is practically impossible. We lack the information to decide which fitted curve is appropriate, and the next data points could lie almost anywhere. This is true in the best of circumstances, where no unexpected development puts a singularity on the time line and makes extrapolation impossible.

Astronomy is as old as human history, and probably older. Mathematics, physics, and chemistry go back at least to Archimedes, in the third century B.C. Biology and medicine certainly predate Hippocrates, who lived around 400 B.C.

Computers, even if we are generous with our chronology, stretch back at most to about 1832, when Charles Babbage began to formulate the ideas for a computing machine—an "analytical engine"—which he was never able to build. In terms of practical experience as to what a general purpose computer can do, we are limited to the half-century since ENIAC began operations in 1946.

With a short history, our curve of projection can go almost anywhere. But computers, and what they may lead to, form so important a part of the human future that they cannot be ignored. We must assess where we are today, and where computers can plausibly go in the next decades and centuries. To see how computers have changed the science fiction world, recall that Heinlein and Clement, to name just two of the field's most famous and scientifically responsible writers, assumed that their heroes would be carrying slide rules around with them in the far future. The slide rule went the way of the dodo in the early 1970s.

How long before our style of personal computer shares history with the slide rule? With that cautionary note in the back of our minds, we begin with some known facts.

In the years since 1946, the speed of electronic computers has increased by a factor of two every two years. There is no evidence that the rate of increase is slowing down. By 2006, machines a billion times as fast as ENIAC will exist.

Between 1946 and 1950, a handful of machines were built. One famous projection, by the management of IBM, estimated that a total of five computers would serve the total needs of the United States. Today, restricting ourselves to general purpose computers, and ignoring the ubiquitous special purpose machines that inhabit everything from thermostats to automobile fuel injection systems, several hundred million machines are in use around the world. Many ordinary households have more than one computer. I had to pause to count, before I realized that here in my house as I write (on a computer, of course) there are seven working machines. It was eight machines until a couple of months ago, when I threw away a computer purchased in 1981 which, although still working, was hopelessly out of date.

Not only speeds and numbers of machines have

changed. Consider storage. The first machine for which I ever wrote a program was called *DEUCE*. It was big enough to walk inside. The engineers would do just that, tapping at suspect vacuum tubes with a screwdriver when the electronics were proving balky. Machine errors were as common a cause of trouble as programing errors; and the latter were dreadfully frequent, because we were working at a level so close to basic machine logic that today the machine would be considered impossible to program.

DEUCE had 402 words of high-speed (mercury delay line) memory, and 8,192 words of back-up (rotating drum) memory. The machine that I just threw away had several times that much storage. There were no tapes or disks on *DEUCE*, only punched cards for input and for intermediate or final output.

Today, it is a poor personal computer that does not have several hundred million bytes of disk storage. Applications programs have expanded to fill the space available. The word processing language I am using requires five million bytes. I'm not sure what it does with them, because I stay several layers of programming languages away from the basic machine instructions.

Costs have been coming down too, to the point where the choice, to buy or to wait, is not an easy one. We know that in a year's time we will receive several times as much capability for today's price. Costs have decreased in about the same way that speeds have gone up. We receive a billion times as much computing for a dollar as we did in 1946.

Where will all this end? There must be limits to speed of performance and memory, but the past half century is no help at all in defining them. We must look elsewhere for the limiting variables.

Physics provides two seeming limits. First, there is a definite relationship between the frequency with which something oscillates, and the associated energy. Specifically, $E = hf$ where E is energy, f is frequency, and h is Planck's constant.

If we are performing 10^{15} operations per second, the associated energy is about one electron volt. Recalling from Chapter 5 that the binding energy of electrons to atoms is just a few electron volts, we see that higher speeds than 10^{15} operations a second will ionize atoms and rip them apart. Here, then, is one apparent limit that we will not be able to surmount.

Actually, we could do better by a factor of a million if we were to build a computer completely from nuclear matter. That would be possible on the surface of a neutron star, but the location might cause other problems.

The speed of light provides a second limit. In order for the computer to function, electrical signals (which travel at light speed) must be able to pass from one component of the computer to another. Since the speed of light is about 300,000 kilometers a second, in 10^{-15} seconds, the time for one operation, a light-speed signal will travel only 0.3 micrometers. A hydrogen atom is about 10^{-4} micrometers across, and all other atoms are bigger; therefore, in 10^{-15} seconds light travels only a distance of at most a couple of hundred atoms. If we could make components that small (currently we cannot) then again we would have a limit of 10^{15} operations a second. A computer made entirely from nuclear matter could push the size limit a million times smaller.

To set this in perspective, there are computers today whose circuit speeds exceed tens of billions of operations a second. We seem, at last, to be able to make a definite statement: computers have a limiting speed of operations which is certainly less than a hundred thousand times what we can achieve today.

But wait. This speed limit assumes that operations are done sequentially, one after another. This is known as serial operation. Many of today's fastest machines employ *parallel* operation, in which many computations are done at the same time. For example, suppose we have to multiply one column of N numbers by another column. Since the N results are independent, with N multiplier units in our computer we can do all N multiplications in the same time as one multiplication.

There are machines today that perform over a hundred thousand operations in parallel. This pushes the potential speed of operation of a machine up by a factor of 10^5. Is there a limit to the degree of parallelism that can be achieved in computation?

It is difficult to see one. We do know that the limit is far higher than 10^5; the human brain, with its 10^{11} neurons, has slow "hardware." The discharge/recharge rate of a neuron is no more than a thousand times a second. But a human (and many animals) can process an image received from the eyes, and recognize a particular individual's face, in less than a tenth of a second. That is less than one hundred "hardware cycles." Our computers require hundreds of millions of hardware cycles to do the same task, and do it less well. We have to conclude that some huge parallelism is at work in both human and animal brains.

Even without considering radically different methods of computing (which we are now about to do), we see no limits to computational speed.

10.2 Biological computers. We now consider two new approaches to achieving high parallelism. Both are at the limit of today's science. Some people would argue that the biological computing, of this section, and the quantum computing of the next, are beyond the limit.

In Chapter 6, we noted that the DNA molecule can be thought of as an information-storing device. That the DNA of our chromosomes happens to carry our own genetic code makes it of special interest to us, which is why we have a human genome mapping project, but in fact, any information at all can be stored in DNA. For example, let us treat the nucleotide bases, A, C, G, and T, as codes for the numbers 0 to 3. Then we can write, using binary notation, A=(00), C=(01), G=(10), T=(11). The long string of a DNA molecule may then be thought of as a general purpose storage device. The computer stores information in binary form, as 0s and 1s. Any sequence of binary digits can be mapped exactly onto a sequence of nucleotide bases, and vice versa. For example, the digit string (11011001101010001111) is equivalent to (TCGCGGGATT). This is an extremely compact form of storage, since a few cubic centimeters can hold as many as 10^{18} DNA molecules. We also have techniques readily available for snipping, joining, and replicating DNA molecules.

So much for storage. This all seems interesting, but not particularly useful, especially since the process of storing and retrieving DNA-stored information is slow compared with electronic storage and retrieval. How can DNA serve as an actual computer, and how can it solve real problems?

The first DNA computer calculation was performed by Leonard Adleman (Adleman, 1994). The problem that he addressed was one in directed graph theory. We will consider here a slightly different and perhaps more familiar application. It is well-known and notorious in operations research and it is called the *Traveling Salesman Problem*.

Suppose that we are given a set of towns, and know the distance between every pair of them. We assume it may not be the same distance both ways, because of bypasses and one-way streets. If a traveling salesman wants to visit each town just once, and return to his starting place, what is the shortest distance that he must travel?

For two towns, the answer is trivial. Similarly for three, where the town-to-town distances define a unique triangle. With four, five, and six towns, the problem becomes a bit more complicated, but it still seems easy. Here, for example, is the case with five towns, Hull, Hornsea, Beverley, Weighton, Driffield, and their associated distance pairs:

	Hull	Hornsea	Beverley	Weighton	Driffield
Hull	0	17	10	15	17
Hornsea	18	0	6	12	20
Beverley	12	5	0	14	19
Weighton	12	11	14	0	7
Driffield	16	21	18	6	0

We read this as telling us that the distance *from* Hull *to* Weighton is 15 miles (first row, fourth column), while the distance *from* Weighton *to* Hull is only 12 miles (fourth row, first column).

There is a simple and absolutely guaranteed way of generating the best solution. Write down the five towns in every possible order, add up the distances for each case, and pick the shortest sum. We can reduce the work somewhat, by noting that it does not matter where we start; so the closed circuit Hull-Beverley-Hornsea-Weighton-Driffield-Hull must give the same total distance as Hornsea-Weighton-Driffield-Hull-Beverley-Hornsea. However, a path may not give the same distance as the reverse path.

Applying the writing-out method to this case, we have to look at $4\times3\times2\times1=24$ possible routes. The salesman's minimum distance is 50 miles, and his route is Hull-Beverley-Hornsea-Weighton-Driffield-Hull.

It may seem hard to believe, but if we specify not five but a hundred towns, the method we have proposed is too much even for today's fastest computers. It is easy to see why. It does not matter where we begin. So pick a town. Then we have 99 choices for the second town; for each of those we have 98 choices for the third town; then 97 choices for the fourth town, and so on. We have to evaluate a total of $99\times98\times97\ldots2\times1$ cases. This number is called factorial 99, and it equals about 10^{156}. The lifetime of the universe does not provide enough computing time to enumerate all the routes.

Of course, direct enumeration is a safe but not a sensible method. The actual approach to tackle the problem would use some variation of a method known as linear programming. Even this, however, will not give us an exact answer. We are forced to go to approximate methods that provide good (but not necessarily best) solutions.

Now let us see how to solve the problem with DNA computation. First, we give to each town a string of 20 DNA nucleotides. The number 20 is long enough to prevent confusion in the calculation. There is no difficulty if we preferred 40, or 100, or any other even number, but 20 is sufficient for this small example.

Here are suitable single DNA strands, where we have

deliberately written each one as two ten-string pieces, and the gaps after the tenth nucleotide are introduced only for convenience of reading:

Hull	TCGCGGGATT	AGACTGTAAG
Hornsea	GTTCGAAGTC	AGTCGTACCT
Beverley	AGCTTATATC	GGTATATGGC
Weighton	ATATGGCGAA	CAGTCGTGCG
Driffield	CGGGATTAGA	TAATCAGGTA

It does not matter what the particular strings look like, only that they be different. They can be chosen at random.

As we know, to every DNA string there corresponds a complementary string, in which the nucleotides (A,T) and (C,G) are interchanged. Together, a string and its complementary string make up a portion of DNA double helix.

The complementary single DNA strands for Hull, Hornsea, Beverley, Weighton, and Driffield, with gaps after the tenth nucleotide introduced for convenience, are:

Hull	AGCGCCCTAA	TCTGACATTC
Hornsea	CAAGCTTCAG	TCAGCATGGA
Beverley	TCGAATATAG	CCATATACCG
Weighton	TATACCGCTT	GTCAGCACGC
Driffield	GCCCTAATCT	ATTAGTCCAT

Now we proceed as follows:

1) We copy the single strands for each town billions of times.

2) For each town pair, we make a 20-element strand using the last 10 elements of the complementary string of the first town, and the first 10 elements of the complementary string of the second town; thus, for Hull to Weighton the string is:

TCTGACATTC   TATACCGCTT

3) We copy these complementary strands billions of times.

4) For each town-to-town pair, say Hull to Weighton,

we take the associated distance (in this case, from the table of distances we see that this distance is 15).

5) Into the middle of each 20-element single strand made for each town pair, insert a line of paired nucleotides equal in length to the *distance* between that pair. For Hull to Weighton, we then have the string:

TCTGACATTC AAAAAAAAAAAAAAA TATACCGCTT
TTTTTTTTTTTTTTTT

The string is a double strand in its middle, a single strand at its ends. Note that the strand corresponding to travel from Weighton to Hull is different than Hull to Weighton.

The strings were written with spaces only to make the structure easier to see. Leaving out the gaps now, we have for Hull to Weighton:

TCTGACATTCAAAAAAAAAAAAAAAATATACCGCTT
TTTTTTTTTTTTTTTT

We make such a DNA strand for travel from every town to every other. The length of any strand will be 20+the distance between the pair (taken in the right order).

In our five-town example, there will be exactly 20 strands, one for each element of the distance table.

6) Now we are ready to start work. We put all our town strands and our town-to-town complementary strands into a beaker, and stir gently.

What happens? As we know, complementary pairs have a natural affinity for each other. Any string, such as AGTC, will seek to match with its complementary string, TCAG, so as to form a double strand. A longer string will have to match base pairs along a greater length before a double strand can form.

Consider the strand for a particular town, say Weighton. Its strand is ATATGGCGAA CAGTCGTGCG. The last ten elements of the strand will seek to match their complementary strand, GTCAGCACGC. Such a sequence will be found at one end of every town-to-town string originating at Weighton, and it will link up to form a double strand over those ten sites. Suppose it links to the Weighton-to-Beverley connecting strand. The result will look like this:

(Weighton to Beverley strand, 34 bases)

```
        GTCAGCACGC  AAAAAAAAAAAAAA  TCGAATATAG
ATATGGCGAA  CAGTCGTGCG  TTTTTTTTTTTTTT  (match Beverley strand)
(Weighton strand)        (distance 14)
```

The far end of the Weighton-to-Beverley strand will have an affinity for the Beverley strand, so it will link up with the first ten sites of that town. The last ten sites of Beverley will in turn seek out the first ten sites of another compatible town-to-town strand, whose last ten sites will link to another town (which could, of course, be Weighton again). The process will go on, adding town after town, unless the process terminates by a string "catching its tail," linking the final site back to the first one to form a closed loop.

Here, for example, is Weighton to Beverley to Hull, with Hull at the right-hand end all set to connect with some other town-to-town strand:

(Weighton to Beverley)

```
        GTCAGCACGC  AAAAAAAAAAAAAA   TCGAATATAG ...
ATATGGCGAA  CAGTCGTGCG  TTTTTTTTTTTTTT  AGCTTATATC ...
    (Weighton)                          (Beverley ...)
```

(Beverley to Hull)

```
... CCATATACCG  AAAAAAAAAAAA  AGCGCCCTAA
... GGTATATGGC  TTTTTTTTTTTT   TCGCGGGATT  AGACTGTAAG
(Beverley continued)                  (Hull)
```

7) In our beaker we have endless billions of different DNA strings. We now sort them out chemically, according to the number of nucleotide bases in each. We can put an upper limit on the length of "interesting" strings. Any string that has more nucleotide base pairs in it than 100 (20 for each of the five towns) plus the sum of the five maximum distances, cannot be the

solution we want. We reject all those strings and retain the others.

We also have a lower limit. Any string less than a hundred bases long cannot contain all the towns, so it can be rejected.

We return only strands of acceptable length to our beaker.

8) Now we unwind each of the double strands into its two separate strings. We want to find the shortest suitable strand, but to be suitable it must include a sequence of bases from each of the five towns (we might find a loop that went from Beverley to Hornsea to Beverley, total 11 miles and therefore total string length 51 bases, but it would be no good).

9) We introduce into our beaker of acceptable-length strands the *complementary* DNA string for each town. We do this for one town at a time. Thus, the complementary string for Hull is AGCGCCCTAA TCTGACATTC.

Anything with the single strand for Hull will attach to this complementary Hull strand. We separate out only strands with such complete Hull double strands. Into this beaker we introduce the complementary DNA string for Hornsea, CAAGCTTCAG TCAGCATGGA. Only DNA strands with the Hornsea DNA single strand will attach, so we can now separate only those strands containing both Hull and Hornsea complete double strands.

In the same way we use the complementary single strands for Beverley, Weighton, and Driffield, to generate strands that must contain complete double DNA strands representing all five towns.

10) Finally, we select out the shortest strands from all those that pass our tests. Such strands visit each town, and they do so with the shortest possible distance. Analysis of those DNA strands will tell us both the town-to-town route and the distance.

This may seem like an awful lot of work to solve a very simple problem, and of course for a small case like this the DNA computer is certainly overkill. With realistic quantities of DNA, we would find not one strand giving the solution, but billions or trillions of them. Also, no one in their right mind would try a simple exhaustive search method on the Traveling Salesman Problem. It is totally impractical for large-scale networks of cities, and different methods are applied. We give it merely to show an example of the technique.

When DNA computing becomes more sophisticated, we should be able to tackle much harder and bigger problems. Soon after Adleman's original paper, Richard Lipton pointed out how a DNA-based computer could address a difficult class of searches known as "satisfaction" problems (Lipton, 1995). Since then, biological computers have been taken seriously as a computational tool with great although unmeasured potential.

10.3 Quantum computers: making a virtue of necessity.
Computers, as they were envisaged originally by Charles Babbage in the first half of the nineteenth century and implemented in the second half of the twentieth, are deterministic machines. This happens to be one of their principal virtues. A calculation, repeated once or a thousand times, will always yield precisely the same answer.

However, as the size of components shrinks toward the molecular and atomic level, indeterminacy inevitably creeps in. The "classical" computer becomes the "quantum" computer, in which quantum effects and uncertainty appear. As we pointed out in Chapter 2, this is an absolutely essential and inescapable consequence of quantum theory, and if the components are small enough there is no way that quantum effects can be either ignored or avoided.

Is there any way we might make a virtue out of necessity, and use quantum effects to improve the performance of a computer? That question has been asked in detail only in the last few years, though Richard Feynman wondered about the possibility in 1985. The answer is astonishing: a "quantum computer" seems to be theoretically possible (none has yet been built), and its performance may permit the solution of problems quite out of reach of a deterministic, classical computer.

The classical computer is built from components that

each have two possible states, which we might label as "on" and "off," or "up" and "down," or 1 and 0. Any number, as we remarked in discussing biological computers, can be written as a string of 1s and 0s; e.g., the decimal number 891,525 is 11011001101010001111 in binary notation. Binary to decimal and decimal to binary conversion is easy for any number whatsoever.

Our quantum computer will use as components individual electrons. Each has two possible spin states, which we label "u" and "d" for up and down. Twenty electrons would then represent the number 891,525 as ddudduuddudduuddddd.

So far we seem to have accomplished nothing. However, recall that according to quantum theory an electron can be in a "mixed state," part u and part d. A mixed state with two components is termed a quantum bit, or *qubit*, to distinguish it from a classical binary digit, or bit. The classical binary digit is either 0 or 1 (u or d). The corresponding qubit is simultaneously 0 and 1 (u and d).

If we know only that the state of each electron is a mixed state, 20 of them—20 qubits—might represent 2^{20} different numbers. If we perform logical operations on the group of electrons, without ever determining their states, then we have performed computations on all possible numbers they might represent. Operations are being performed in parallel, and to do the same thing with a classical computer we would need 2^{20} processing units—more than a million of them.

The choice of a string of 20 electrons was arbitrary. We could have as easily chosen 100 qubits, or 1,000. That is still a tiny set, compared with the number of electrons in any electric signal. However, $2^{1,000}$ is about 10^{300}. We have a possible parallel operation at a near-incomprehensible level.

The principles described here are clear. Some possible practical applications are already known. For example, the parallel processing provided by qubits can be used to decompose numbers into their prime factors. This is a difficult problem, which with classical computers cannot be solved in practice for very large numbers. The computation time becomes too enormous.

However, we have skated over some of the difficulties involved with quantum computers. The worst one is the extreme sensitivity of a quantum computer to its surroundings. No computer can be completely isolated from the rest of the universe, and tiny interactions will disturb the mixed states needed for the qubits. This is termed the "problem of decoherence," and its practical effect is that any problem solved on a quantum computer must be completed before interaction of the environment causes decoherence.

Quantum computers are very much on the science frontier. Their development stage today may be like that of classical computers in the mid-1940s, when some of the world's smartest people doubted that a reliable electronic computer would ever be built—the failure rate of vacuum tubes was too high. Today, transistors and integrated circuits are so reliable that a hardware error is the last place to look when a program fails (the first place we look is in the computer program code, generated by that quirky and unreliable computer, the human brain).

10.4 Where are the robots? Science fiction writers did a poor job predicting the arrival of near-ubiquitous general purpose computers. What science fiction did predict was robots, mechanical marvels capable of performing all manner of tasks normally associated with humans.

Robots came into science fiction three-quarters of a century ago, in Karel Capek's play *R.U.R.* (Capek, 1920). They have been a staple element of science fiction ever since. In the real world, robots have fared less well. Either they have been confined to the role of robot arms at a fixed location, performing a few limited operations on an assembly line; or they have been slow-moving, clumsy morons, trundling their way with difficulty across a simplified room environment to pick up colored blocks with less skill and accuracy than the average two-year-old.

What went wrong? And when, if ever, will it go right?

The big problem seems to be human hubris. We, aware

of the big and complex brains that set us apart from every other animal, overemphasize the importance of logical thought. At the same time, we tend to diminish the importance of the functions that we share with animals: easy recognition and understanding of environment, easy grasping of objects, effortless locomotion across difficult terrain.

But seeing and walking have a billion years of development effort behind them. We do them well, not because they are intrinsically simple, but because evolution has weeded out anything that found them difficult. We don't even have to think about seeing. Logical thought, on the other hand, has been around for no more than a million years. No wonder we still have trouble doing it. We are proud of our ability, but a fully evolved creature would find thought as effortless, and be as unaware of the complexity of operations that lay behind it, as taking a drink of water.

Recognizing the truth does not solve the problem, but it allows us to place emphasis in the appropriate area. For many years, the "difficult" part of making a robot was assumed to be the logical operations. This led to computer programs that play a near-perfect game of checkers and a powerful game of chess. The hardware/software combination known as Deep Blue beat world-champion Kasparov in 1997, though human fatigue and stress were also factors. At the same time, the program was as helpless as a baby when it came to picking up a chess piece and executing a move. Those functions were performed by Deep Blue's human handlers.

So when will we have a "real" robot, one able to perform useful tasks in the relatively complicated environment of the average home?

The answer is: when development from two directions meet.

Those two directions are:

1) "Top-down" activities, usually referred to as Artificial Intelligence, or just AI, that seek to copy human thought processes as they exist today. AI, after a promising start in the 1960s, stumbled and slowed. One problem is that we don't know exactly what human thought processes

are. As Marvin Minsky has pointed out (Minsky, 1986), the easy part is modeling the activities of the conscious mind. The hard part is the unconscious mind, inaccessible to us and difficult to define and imitate.

2) "Bottom-up" activities, that start with the basic problems of perception and mobility, without introducing the idea of thought at all. This is an "evolutionary" approach, building computers that incrementally model the behavior of animals with complex nervous systems. We know that this can be done, because it happened once already, in Nature. However, we hope to beat Nature's implementation schedule of a few billion years.

When top-down and bottom-up meet, in what Hans Moravec refers to as the "metaphorical golden spike" of robotics (Moravec, 1988), we will have a reasoning computer program (or, more likely, a large interconnected set of programs) with a good associated "lower-level" perception and movement capability. In other words, robots as science fiction has known them for many years.

When?

Moravec says in fifty years or so. He is perhaps not entirely impartial, but if we do not accept the estimates of leaders in the robotics field, whom do we believe? If you introduce working household robots into a story set in 2050, at least some of today's robotics specialists will offer you moral support.

In making his estimate, Moravec relies on two things. First, that the projections quoted at the beginning of this chapter on computer speed, size, and costs are correct. Advances in biological or quantum computers can only serve to bring the date of practical robots closer.

Second, Moravec believes that when the necessary components come together, they will do so very quickly.

We have to ask the question: What next? What will come after reasoning computer programs?

The optimists see a wonderful new partnership, with humans and the machines that they have created moving together into a future where human manual labor is unknown, while mental activities become a splendid joint endeavor.

The pessimists point out that computers are only half

a century old. In another one or two hundred years they may be able to design improved versions of themselves. At that point humans will have served their evolutionary purpose, as a transition stage on the way to a higher life-form. We can bow out, while computers and their descendants inherit the universe. With luck, maybe a few of us will be kept around as historical curiosities.

All of this presumes that the development we describe next does not achieve the potential that many people foresee for it.

10.5 Nanotechnology: the anything machine. Richard Feynman, who is apt to pop up anywhere in the physics of the second half of this century, gave in 1959 a speech that many of his listeners regarded as some kind of joke. It has since come to seem highly prophetic. Feynman noted that whereas the machines we build are all different to a greater or less degree, every electron is identical, as is every proton and every neutron. He suggested that if we built machines one atom at a time, they could be absolutely identical. He also wondered just how small a machine might be made. Could there be electric motors, a millimeter across? If so, then how about a micrometer across? Bacteria are no bigger, and they seem much more complicated than the relatively simple machines that we use in our everyday world.

Suppose that such minute machines can be built, hardly bigger than a large molecule; further, suppose that they can be made *self-replicating*, able, like bacteria, to make endless copies of themselves from raw materials in their environment. Finally, suppose that the machines can be programmed, to perform cooperatively any task we choose for them. At a larger scale, Nature has again beaten us to it. The social insects (ants, bees, termites) form a highly cooperative group of individually simple entities, able in combination to accomplish the complex tasks of colony maintenance and reproduction.

These ideas of tiny, self-replicating, programmable machines were all put together by Eric Drexler, and a name given to the whole concept: *nanotechnology*. In a book, *The Engines of Creation* (Drexler, 1986), and in

subsequent works, he outlined what myriads of these
programmable self-replicating machines might accomplish.

The list includes flawless production with built-in
quality control (misplaced atoms can be detected and
replaced or moved); materials with a strength-to-density
ratio an order of magnitude better than anything we
have today (useful for the exotic space applications of
Chapter 8); molecular-level disease diagnosis and tissue repair, of a non-intrusive nature—in other words,
we would be unaware of the presence of the machines
within our own bodies; and "smart" home service and
transportation systems, capable of automatic self-
diagnosis and component replacement. When suitable
programs have been developed—once only, for each
application—all of these things will be available for the
price of the raw materials.

Putting the potential applications together, we seem
to have an anything machine. Any item that we can
define completely can be built, inexpensively, in large
quantities, provided only that the basic materials are
inexpensive. Spaceships and aircraft will grow themselves.
Household chef units will develop the meals of our choice
from basic non-biotic components. Our bodies will have
their own built-in health maintenance systems. Build-up
of arterial plaque or cholesterol will be prevented; eyes
or ears that do not perform perfectly will be modified;
digestion will become a monitored process, digestive
disorders a thing of the past. We will all enjoy perfect
health and a prolonged life expectancy. Perhaps, if the
nanomachines are much smaller than the cell level, and
can work on our telomeres, we will have the potential
to live forever.

There is, naturally, a dark potential to all this. What
happens if the self-replicating machines go out of control?
In *Cold as Ice*, I introduced Fishel's Law and Epitaph:
*Smart is dumb; it is unwise to build too much intelligence
into a self-replicating machine.* Greg Bear saw the total
end of humanity, when the nanomachines of *Blood Music*
took over (Bear, 1985).

We cannot say whether a world with fully-developed
nanotechnology will be good or bad; what we can say

is that it cannot be predicted from today's world. Nanotechnology represents a singularity of the time-line, a point beyond which rational extrapolation is impossible.

10.6 Artificial life and assisted thought. In 1969, the English mathematician John Horton Conway introduced a paper-and-pencil game with an appropriate name: *Life*. Given a large sheet of paper marked into small squares, together with a few simple rules, objects can be generated that seem to possess many of the attributes of living organisms. They move across the page, grow, reproduce copies of themselves and of certain other organisms, and kill other organisms by "eating" them. The game was a big success in academic circles, and computer versions of it soon appeared.

Slowly, through the 1970s and 1980s, the realization grew that computers might also be useful in studying life with a small "l." The behavior of competing organisms could be modeled. Those organisms could then be "released" into a computer environment and allowed to "evolve." The results would provide valuable information about population dynamics.

But must it stop there? If we take the science fictional next step, we already pointed out in discussing biological computers that the DNA of any organism can be put into exact correspondence with a string of binary digits. In a real sense, those digits represent the organism. Given the digit string and suitable technology, we could construct the DNA, introduce it into the superstructure of a cell, and grow the organism itself.

We could do that, but why should we bother? We know that computer circuits operate millions of times faster than our own nerve cells. Couldn't we take various kinds of DNA, representing different organisms, prescribe the rules within the computer for growing the organism, and let the competition for genetic survival run not in the real world, but inside the computer? Maybe we can in that way speed up the process of evolution by a factor of many millions, and watch the emergence of new species in close to real time.

The practical problems are enormous. There is so much

that we do not know about the development of a living creature. The necessary data must certainly be in the DNA, to allow an eye and a kidney and a brain to develop from a single original cell, but we have little idea how this "cellular differentiation" takes place. In fact, it is one of the central mysteries of biology.

Meanwhile, we look at another possibility. Suppose, as discussed in the previous chapter, we were able to download into a computer the information content of a human brain. If Roger Penrose is right (See Chapter 13), this may not be possible until we have a quantum computer able to match the quantum functions of the human brain; but let us suppose we have that. Now we have something that does not evolve, as DNA representations might evolve, but thinks in the virtual environment provided inside the computer. This is artificial life, of a specific and peculiar kind. For one thing, the speed of computer thought should eventually exceed the speed of our flesh-and-blood wetware by a factor of millions or billions.

Take this one step further. The size of computers constantly gets smaller and smaller. Already, we could tuck a powerful machine away into a little spare head room, maybe in one of the *sulci* of the brain or in a sinus cavity. Even if the machine lacked the full thinking capacity of a human brain, it could certainly perform the routine mathematical functions which for most of us are anything but routine: lengthy arithmetic, detailed logical analyses, symbol manipulation, and compound probabilities. (Maybe we should add elementary arithmetic to the list, since the ability to do mental calculations, such as computing change from twenty dollars, seems to be waning fast.)

Nature took a billion years to provide us with brains able to perceive, manipulate, and move our bodies with effortless ease. Maybe we can give ourselves powers of effortless thought, to go with those other desirable features, in a century or two.

CHAPTER 11
Chaos: The Unlicked Bear-Whelp

My original plan was to leave this chapter out of the book, as too technical. However, it was suggested to me that the science of chaos theory can be a fertile source of stories; more than that, it was pointed out that a story ("Feigenbaum Number," Kress, 1995) had been written drawing explicitly on an earlier article of mine on the subject. Faced with such direct evidence, I changed my mind. I did, however, remove most of the equations. I hope the result is still intelligible.

11.1 Chaos: new, or old? The Greek word "chaos" referred to the formless or disordered state before the beginning of the universe. The word has also been a part of the English language for a long time. Thus in Shakespeare's *Henry VI*, Part Three, the Duke of Gloucester (who in the next play of the series will become King Richard III, and romp about the stage in unabashed villainy) is complaining about his physical deformities. He is, he says, "like to a Chaos, or an unlick'd bear-whelp, that carries no impression like the dam." *Chaos*: something essentially random, an object or being without a defined shape.

Those lines were written about 1590. The idea of chaos is old, but *chaos theory* is a new term. Twenty years ago, no popular article had ever been written containing that expression. Ten years ago, the subject was all the rage. It was hard to find a science magazine *without* finding an article on chaos theory, complete with stunning color illustrations. Today, the fervor has faded, but the state of the subject is still unclear (perhaps appropriate, for something called *chaos theory*). Most of the articles seeking to explain what it is about are even less clear.

Part of the problem is newness. When someone writes about, say, quantum theory, the subject has to be presented as difficult, and subtle, and mysterious, because it *is* difficult, and subtle, and mysterious. To describe it any other way would be simply misleading. In the past sixty years, however, the mysteries have had time to become old friends of the professionals in the field. There are certainly enigmas, logical whirlpools into which you can fall and never get out, but at least the *locations* of those trouble spots are known. Writing about any well-established subject such as quantum theory is therefore in some sense easy.

In the case of chaos theory, by contrast, *everything* is new and fragmented; we face the other extreme. We are adrift on an ocean of uncertainties, guided by partial and inadequate maps, and it is too soon to know where the central mysteries of the subject reside.

Or, worse yet, to know if those mysteries are worth taking the time to explore. *Is* chaos a real "theory," something which will change the scientific world in a basic way, as that world was changed by Newtonian mechanics, quantum theory, and relativity? Or is it something essentially trivial, a subject which at the moment is benefiting from a catchy name and so enjoying a certain glamour, as in the past there have been fads for orgone theory, mesmerism, dianetics, and pyramidology?

I will defer consideration of that question, until we have had a look at the bases of chaos theory, where it came from, and where it seems to lead us. Then we can come back to examine its long-term prospects.

11.2 How to become famous. One excellent way to make a great scientific discovery is to take a fact that everyone knows must be the case—because "common sense demands it"—and ask what would happen if it were not true.

For example, it is obvious that the Earth is fixed. It *has* to be standing still, because it feels as though it is standing still. The Sun moves around it. Copernicus, by suggesting that the Earth revolves around the Sun, made the fundamental break with medieval thinking and set in train the whole of modern astronomy.

Similarly, it was clear to the ancients that unless you keep on pushing a moving object, it will slow down and stop. By taking the contrary view, that it takes a force (such as friction with the ground, or air resistance) to *stop* something, and otherwise it would just keep going, Galileo and Newton created modern mechanics.

Another case: To most people living before 1850, there was no question that animal and plant species are all so well-defined and different from each other that they must have been created, type by type, at some distinct time in the past. Charles Darwin and Alfred Russel Wallace, in suggesting in the 1850s a mechanism by which one form could *change* over time to another in response to natural environmental pressures, allowed a very different world view to develop. The theory of evolution and natural selection permitted species to be regarded as fluid entities, constantly changing, and all ultimately derived from the simplest of primeval life forms.

And, to take one more example, it was clear to everyone before 1900 that if you kept on accelerating an object, by applying force to it, it would move faster and faster until it was finally traveling faster than light. By taking the speed of light as an upper limit to possible speeds, and requiring that this speed to be the same for all observers, Einstein was led to formulate the theory of relativity.

It may make you famous, but it is a risky business, this offering of scientific theories that ask people to abandon their long-cherished beliefs about what "just must be so." As Thomas Huxley remarked, it is the customary fate of new truths to begin as heresies.

Huxley was speaking metaphorically, but a few hundred years ago he could have been speaking literally.

Copernicus did not allow his work on the movement of the Earth around the Sun to be published in full until 1543, when he was on his deathbed, nearly 30 years after he had first developed the ideas. He probably did the right thing. Fifty-seven years later Giordano Bruno was gagged and burned at the stake for proposing ideas in conflict with theology, namely, that the universe is infinite and there are many populated worlds. Thirty-three years after that, Galileo was made to appear before the Inquisition and threatened with torture because of his "heretical" ideas. His work remained on the Catholic Church's Index of prohibited books for over two hundred years.

By the nineteenth century critics could no longer have a scientist burned at the stake, even though they may have wanted to. Darwin was merely denounced as a tool of Satan. However, anyone who thinks this issue is over and done with can go today and have a good argument about evolution and natural selection with the numerous idiots who proclaim themselves to be scientific creationists.

Albert Einstein fared better, mainly because most people had no idea what he was talking about. However, from 1905 to his death in 1955 he became the target of every crank and scientific nitwit outside (and often inside) the lunatic asylums.

Today we will be discussing an idea, contrary to common sense, that has been developing in the past twenty years. So far its proposers have escaped extreme censure, though in the early days their careers may have suffered because no one believed them—or understood what they were talking about.

11.3 Building models. The idea at the heart of chaos theory can be simply stated, but we will have to wind our way into it.

Five hundred years ago, mathematics was considered essential for bookkeeping, surveying, and trading, but it was not considered to have much to do with the physical processes of Nature. Why should it? What do abstract symbols on a piece of paper have to do with the movement of the planets, the flow of rivers, the blowing of soap bubbles, the flight of kites, or the design of buildings?

Little by little, that view changed. Scientists found that physical processes could be described by equations, and solving those equations allowed predictions to be made about the real world. More to the point, they were *correct* predictions. By the nineteenth century, the fact that manipulation of the purely abstract entities of mathematics could somehow tell us how the real world would behave was no longer a surprise. Sir James Jeans could happily state, in 1930, "*all* the pictures which science now draws of nature, and which alone seem capable of according with observational fact, are *mathematical* pictures," and " . . . the universe appears to have been designed by a pure mathematician."

The mystery had vanished, or been subsumed into divinity. But it should not have. It is a mystery still.

I would like to illustrate this point with the simplest problem of Newtonian mechanics. Suppose that we have an object moving along a line with a constant acceleration. It is easy to set up a situation in the real world in which an object so moves, at least approximately.

It is also easy to describe this situation mathematically, and to determine how the final position depends on the speed and initial position. When we do this, we find that a tiny change in initial speed or position causes a small change in final speed and position. We say that the solution is a continuous function of the input variables.

This is an especially simple example, but scientists are at ease with far more complex cases.

Do you want to know how a fluid will move? Write down a rather complex equation (to be specific, the three-dimensional time-dependent Navier-Stokes equation for compressible, viscous flow). Solve the equation. That's not a simple proposition, and you may have to resort to a computer. But when you have the results, you expect them to apply to real fluids. If they do not, it is because the equation you began with was not quite right—maybe we need to worry about electromagnetic forces, or plasma effects. Or maybe the integration method you used was numerically unstable, or the finite difference interval too crude. The idea that the mathematics cannot describe the physical world never even occurs to most scientists. They have in the back of their minds an idea first made explicit by Laplace: the whole universe is calculable, by defined mathematical laws. Laplace said that if you told him (or rather, if you told a demon, who was capable of taking in all the information) the position and speed of every particle in the Universe, at one moment, he would be able to define the Universe's entire future, and also its whole past.

The twentieth century, and the introduction by Heisenberg of the Uncertainty Principle, weakened that statement, because it showed that it was impossible to know precisely the position and speed of a body. Nonetheless, the principle that mathematics can *exactly model reality* is usually still unquestioned.

It should be, because it is absolutely extraordinary that the pencil and paper scrawls that we make in our studies correspond to activities in the real world outside.

Now, hidden away in the assumption that the world can be described by mathematics there is another one; one so subtle that most people never gave it a thought. This is the assumption that chaos theory makes explicit, and then challenges. We state it as follows:

Simple equations must have simple solutions.

There is no reason why this should be so, except that it seems that common sense demands it. And, of course, we have not defined "simple."

Let us return to our accelerating object, where we have a simple-seeming equation, and an explicit solution. One requirement of a simple solution is that it should not "jump around" when we make a very small change in the system it describes. For example, if we consider two cases of an accelerated object, and the only difference between them is a tiny change in the original position of the object, we would expect a small change in the *final* position. And this is the case. That is exactly what was meant by the earlier statement, that the solution was a continuous function of the inputs.

But now consider another simple physical system, a

rigid pendulum (this was one of the first cases where the ideas of chaos theory emerged). If we give the pendulum a small push, it swings back and forward. Push it a little harder, and a little harder, and what happens? Well, for a while it makes bigger and bigger swings. But at some point, a very small change to the push causes a totally different type of motion. Instead of swinging back and forward, the pendulum keeps on going, right over the top and down the other side. If we write the expression for the angle as a function of time, in one case the angle is a *periodic* function (back and forth) and in the other case it is constantly increasing (round and round). And the change from one to the other occurs when we make an *infinitesimal* change in the initial speed of the pendulum bob. This type of behavior is known as a *bifurcation* in the behavior of the solution, and it is a worrying thing. A simple equation begins to exhibit a complicated solution. The solution of the problem is no longer a continuous function of the input variables.

At this point, the reasonable reaction might well be, so what? All that we have done is show that certain simple equations don't have really simple solutions. That does not seem like an earth-shaking discovery. For one thing, the boundary between the two types of solution for the pendulum, oscillating and rotating, is quite clear-cut. It is not as though the definition of the location of the boundary itself were a problem.

Can situations arise where this *is* a problem? Where the boundary is difficult to define in an intuitive way? The answer is, yes. In the next section we will consider simple systems that give rise to highly complicated boundaries between regions of fundamentally different behavior.

11.4 Iterated functions. Some people have a built-in mistrust of anything that involves the calculus. When you use it in any sort of argument, they say, logic and clarity have already departed. The solutions for examples I have given so far implied that we write down and solve a differential equation, so calculus was needed to define the behavior of the solutions. However, we don't need calculus to demonstrate fundamentally chaotic behavior; and many of the first explorations of what we now think of as chaotic functions were done without calculus. They employed what is called *iterated function theory*. Despite an imposing name, the fundamentals of iterated function theory are so simple that they can be done with an absolute minimum knowledge of mathematics. They do, however, benefit from the assistance of computers, since they call for large amounts of tedious computation.

Consider the following very simple operation. Take two numbers, x and r. Form the value $y = rx(1-x)$.

Now plug the value of y back in as a new value for x. Repeat this process, over and over.

For example, suppose that we take r=2, and start with x=0.1. Then we find y=0.18.

Plug that value in as a new value for x, still using r=2, and we find a new value, y=0.2952.

Keep going, to find a sequence of y's, 0.18, 0.2952, 0.4161, 0.4859, 0.4996, 0.5000, 0.5000 . . .

In the language of mathematics, the sequence of y's has *converged* to the value 0.5. Moreover, for any starting value of x, between 0 and 1, we will always converge to the same value, 0.5, for r=2.

Here is the sequence when we begin with x = 0.6: 0.4800, 0.4992, 0.5000, 0.5000 . . .

Because the final value of y does not depend on the starting value, it is termed an *attractor* for this system, since it "draws in" any sequence to itself.

The value of the attractor depends on r. If we start with some other value of r, say r=2.5, we still produce a convergent sequence. For example, if for r=2.5 we begin with x = 0.1, we find successive values: 0.225, 0.4359, 0.6147, 0.5921, 0.6038, 0.5981, . . . 0.6. Starting with a different x still gives the same final value, 0.6.

For anyone who is familiar with a programming language such as C or even BASIC (Have you noticed how computers are used less and less to *compute*?), I recommend playing this game for yourself. The whole program is only a dozen lines long. *Suggestion*: Run the program in double precision, so you don't get trouble with round-off errors. *Warning*: Larking around with this sort of thing will consume hours and hours of your time.

The situation does not change significantly with r=3. We find the sequence of values: 0.2700, 0.5913, 0.7250, 0.5981, 0.7211 . . . 0.6667. This time it takes thousands of iterations to get to a final converged value, but it makes it there in the end. Even after only a dozen or two iterations we can begin to see it "settling-in" to its final value.

There have been no surprises so far. What happens if we increase r a bit more, to 3.1? We might expect that we will converge, but even more slowly, to a single final value.

We would be wrong. Something very odd happens. The sequence of numbers that we generate has a regular structure, but now the values alternate between two different numbers, 0.7645, and 0.5580. *Both* these are attractors for the sequence. It is as though the sequence cannot make up its mind. When r is increased past the value 3, the sequence "splits" to two permitted values, which we will call "states," and these occur alternately.

Let us increase the value of r again, to 3.4. We find the same behavior, a sequence that alternates between two values.

But by r=3.5, things have changed again. The sequence has *four* states, four values that repeat one after the other. For r=3.5, we find the final sequence values: 0.3828, 0.5009, 0.8269, and 0.8750. Again, it does not matter what value of x we started with, we will always converge on those same four attractors.

Let us pause for a moment and put on our mathematical hats. If a mathematician is asked the question, Does the iteration y=rx(1−x) converge to a final value?, he will proceed as follows:

Suppose that there is a final converged value, V, towards which the iteration converges. Then when we reach that value, no matter how many iterations it takes, at the final step x will be equal to V, and so will y. Thus we must have V=rV(1−V).

Solving for V, we find V=0, which is a legitimate but uninteresting solution, or V=(r−1)/r. This single value will apply, no matter how big r may be. For example, if r=2.5, then V=1.5/2.5=0.6, which is what we found. Similarly, for r=3.5, we calculate V=2.5/3.5=0.7142857.

But this is not what we found when we did the actual iteration. We did not converge to that value at all, but instead we obtained a set of four values that cycled among themselves. So let us ask the question, what would happen if we *began* with x=0.7142857, as our starting guess? We certainly have the right to use any initial value that we choose. Surely, the value would simply stay there?

No, it would not.

What we would find is that on each iteration, the value of y changes. It remains close to 0.7142857 on the first few calculations, then it—quite quickly—diverges from that value and homes in on the four values that we just mentioned: 0.3828, 0.5009, etc. In mathematical terms, the value 0.7142857 is a solution of the iterative process for r=3.5. But it is an *unstable* solution. If we start there, we will rapidly move away to other multiple values.

Let us return to the iterative process. By now we are not sure what will happen when we increase r. But we can begin to make some guesses. Bigger values of r seem to lead to more and more different values, among which the sequence will oscillate, and it seems though the number of these values will always be a power of two. Furthermore, the "splitting points" seem to be coming faster and faster.

Take r=3.52, or 3.53, or 3.54. We still have four values that alternate. But by r=3.55, things have changed again. We now find *eight* different values that repeat, one after the other. By r=3.565, we have 16 different values that occur in a fixed order, over and over, as we compute the next elements of the sequence.

It is pretty clear that we are approaching some sort of crisis, since the increments that we can make in r, without changing the nature of the sequence, are getting smaller and smaller. In fact, the critical value of r is known to many significant figures. It is r=3.569945668. . . . As we approach that value there are 2^n states in the sequence, and n is growing fast.

What happens if we take r *bigger* than this, say r=3.7? We still produce a sequence—there is no difficulty at all with the computations—but it is a sequence without any sign of regularity. There are no attractors, and all values seem equally likely. It is fair to say that it is *chaos*, and the region beyond the critical value of r is often called the *chaos regime*.

This may look like a very special case, because all the calculations were done based on one particular function, $y=rx(1-x)$. However, it turns out that the choice of function is much less important than one would expect. If we substituted any up-and-down curve between zero and one we would get a similar result. As r increases, the curve "splits" again and again. There is a value of r for which the behavior becomes chaotic.

For example, suppose that we use the form $y=r\cdot\sin(x)/4$ (the factor of 4 is to make sure that the maximum value of y is the same as in the first case, namely, 1/4). By the time we reach r=3.4 we have four different values repeating in the sequence. For r=3.45 we have eight attractors. Strangest of all, the way in which we approach the critical value for this function has much in common with the way we approached it for the first function that we used. They both depend on a single convergence number that tells the rate at which new states will be introduced as r is increased. That convergence number is 4.669201609 . . . , and is known as the *Feigenbaum number*, after Mitchell Feigenbaum, who first explored in detail this property of iterated sequences. This property of common convergence behavior, independent of the particular function used for the iteration, is called *universality*. It seems a little presumptuous as a name, but maybe it won't, in twenty years time.

This discussion of iterated functions may strike you as rather tedious, very complicated, very specialized, and a way of obtaining very little for a great deal of work. However, the right way to view what we have just done is this: we have found a critical value, less than which there is a predictable, although increasingly complicated

behavior, and above which there is a completely different
and chaotic behavior. Moreover, as we approach the
critical value, the number of possible states of the system
increases very rapidly, and tends to infinity.

To anyone who has done work in the field of fluid
dynamics, that is a very suggestive result. For fluid flow
there is a critical value below which the fluid motion
is totally smooth and predictable (laminar flow) and
above which it is totally unpredictable and chaotic
(turbulent flow). Purists will object to my characterizing
turbulence as "chaotic," since although it appears chaotic
and disorganized as a whole, there is a great deal of
structure on the small scale since millions of molecules
must move together in an organized way. However, the
number of states in turbulent flow is infinite, and there
has been much discussion of the way in which the single
state of laminar flow changes to the many states of
turbulent flow. Landau proposed that the new states must
come into being one at a time. It was also assumed that
turbulent behavior arose as a consequence of the very
complicated equations of fluid dynamics.

Remember the "common sense rule": *Simple equations
must have simple solutions*. And therefore, complicated
behavior should only arise from complicated equations.
For the first time, we see that this may be wrong. A very
simple system is exhibiting very complicated behavior,
reminiscent of what happens with fluid flow. Depending
on some critical variable, it may appear totally predictable
and well-behaved, or totally unpredictable and chaotic.
Moreover, experiments show that in turbulence the new,
disorganized states come into being not one by one, but
*through a doubling process as the critical parameter is
approached*. Maybe turbulence is a consequence of something in the fluid flow equations that is unrelated to their
complexity—a hidden structure that is present even in such
simple equations as we have been studying.

This iterated function game is interesting, even suggestive, but to a physicist it was for a long time little
more than that. Physics does not deal with computer
games, went the argument. It deals with mathematical
models that describe a physical system, in a majority

of cases through a series of differential equations. These equations are solved, to build an idea of how Nature will behave in any given circumstance.

The trouble is, although such an approach works wonderfully well in many cases, there are classes of problems that it doesn't seem to touch. Turbulence is one. "Simple" systems, like the dripping of water from a faucet, can be modeled in principle, but in practice the difficulties in formulation and solution are so tremendous that no one has ever offered a working analysis of a dripping tap.

The problems where the classical approach breaks down often have one thing in common: they involve a random, or apparently random, element. Water in a stream breaks around a stone this way, then that way. A snowflake forms from supersaturated vapor, and every one is different. A tap drips, then does not drip, in an apparently random way. All these problems are described by quite different systems of equations. What scientists wanted to see was *physical problems*, described by good old differential equations, that also displayed bifurcations, and universality, and chaotic behavior.

They had isolated examples already. For example, the chemical systems that rejoice in the names of the Belousov-Zhabotinsky reaction and the Brusselator exhibit a two-state cyclic behavior. So does the life cycle of the slime mold, *Dictyostelium discoideum*. However, such systems are very tricky to study for the occurrence of such things as bifurcations, and involve all the messiness of real-world experiments. Iterated function theory was something that could be explored in the precise and austere world of computer logic, unhindered by the intrusion of the external world.

We must get to that external and real world eventually, but before we do so, let's take a look at another element of iterated function theory. This one has become very famous in its own right (rather more so, in my opinion, than it deserves to be for its physical significance, but perhaps justifiably most famous for its artistic significance).

The subject is *fractals*, and the contribution to art is called the Mandelbrot Set.

11.5 Sick curves and fractals. Compare the system we have just been studying with the case of the pendulum. There we had a critical *curve*, rather than a critical value. On the other hand, the behavior on both sides of the critical curve was not chaotic. Also, the curve itself was well-behaved, meaning that it was "smooth" and predictable in its shape.

Is there a simple system that on the one hand exhibits a critical *curve*, and on the other hand shows chaotic behavior?

There is. It is one studied in detail by Benoit Mandelbrot, and it gives rise to a series of amazing objects (one hesitates to call them curves, or areas).

We just looked at a case of an iterated function where only one variable was involved. We used x to compute y, then replaced x with y, and calculated a new y, and so on. It is no more difficult to do this, at least in principle, if there are two starting values, used to compute two new values. For example, we could have:

$$y = (w^2 - x^2) + a$$
$$z = 2wx + b$$

and when we had computed a pair (y,z) we could use them to replace the pair (w,x). (Readers familiar with complex variable theory will see that I am simply writing the relation $z = z^2 + c$, where z and c are complex numbers, in a less elegant form.)

What happens if we take a pair of constants, (a,b), plug in zero starting values for w and x, and let our computers run out lots of pairs, (y,z)? This is a kind of two-dimensional equivalent to what we did with the function $y = rx(1-x)$, and we might think that we will find similar behavior, with a critical *curve* replacing the critical value.

What happens is much more surprising. We can plot our (y,z) values in two dimensions, just as we could plot speeds and positions for the case of the pendulum to make a *phase space diagram*. And, just as was the case with the pendulum, we will find that the whole plane divides into separate regions, with boundaries between them. The boundaries are the boundary curves of the "Mandelbrot set," as it is called. If, when we start with an (a,b) pair and iterate for (y,z) values, one or both of y and z run off towards infinity, then the point (a,b) is *not* a member of the Mandelbrot set. If the (y,z) pairs settle down to some value, or if they cycle around a series of values without ever diverging off towards infinity, then the point (a,b) is a member of the Mandelbrot set. The tricky case is for points on the boundary, since convergence is slowest there for the (y,z) sequence. However, those boundaries can be mapped. And they are as far as can be imagined from the simple, well-behaved curve that divided the two types of behavior of the pendulum. Instead of being smooth, they are intensely spiky; instead of just one curve, there is an infinite number.

The results of plotting the Mandelbrot set can be found in many articles, because they have a strange beauty unlike anything else in mathematics. Rather than drawing them here, I will refer you to James Gleick's book, *Chaos: Making a New Science* (Gleick, 1987), which shows some beautiful color examples of parts of the set. All this, remember, comes from the simple function we defined, iterated over and over to produce pairs of (y,z) values corresponding to a particular choice of a and b. The colors seen in so many art shows, by the way, while not exactly a cheat, are not fundamental to the Mandelbrot set itself. They are assigned depending on how many iterations it takes to bring the (y,z) values to convergence, or to a stable repeating pattern.

The Mandelbrot set also exhibits a feature known as *scaling*, which is very important in many areas of physics. It says, in its simplest terms, that you cannot tell the absolute scale of the phenomenon you are examining from the structure of the phenomenon itself.

That needs some explanation. Suppose that you want to know the size of a given object—say, a snowflake. One absolute measure, although a rather difficult one to put into practice, would be to count the number of atoms in that snowflake. Atoms are fundamental units, and they do not change in their size.

But suppose that instead of the number of atoms, you tried to use a different measure, say, the total *area* of the snowflake. That sounds much easier than looking at the individual atoms. But you would run into a problem, because as you look at the surface of the snowflake more and more closely, it becomes more and more detailed. A little piece of a snowflake has a surface that looks very much like a little piece of a little piece of a snowflake; a little piece of a little piece resembles a little piece of a little piece of a little piece, and so on. It stays that way until you are actually seeing the atoms. Then you at last have the basis for an absolute scale.

Mathematical entities, unlike snowflakes, are not made up of atoms. There are many mathematical objects that "scale forever," meaning that each level of more detailed structure resembles the one before it. The observer has no way of assigning any absolute scale to the structure. The sequence-doubling phenomenon that we looked at earlier is rather like that. There is a constant ratio between the distances at which the doublings take place, and that information alone is not enough to tell you how close you are to the critical value in absolute terms.

Similarly, by examining a single piece of the Mandelbrot set it is impossible to tell at what level of detail the set is being examined. The set can be examined more and more closely, forever, and simply continues to exhibit more and more detail. There is never a place where we arrive at the individual "atoms" that make up the set. In this respect, the set differs from anything encountered in nature, where the fundamental particles provide a final absolute scaling. Even so, there are in nature things that exhibit scaling over many orders of magnitude. One of the most famous examples is a coastline. If you ask "How long is the coastline of the United States?" a first thought is that you can go to a map and measure it. Then it's obvious that the map has smoothed the real coastline. You need to go to larger scale maps, and larger scale maps. A coastline "scales," like the surface of a snowflake, all the way down to the individual rocks and grains of sand. You find larger and larger numbers for the length of the coast. Another natural phenomenon that exhibits scaling is—significantly—turbulent flow. Ripples ride on whirls that ride on vortices that sit on swirls that are made up of eddies, on and on.

There are classes of mathematical curves that, like coastlines, do not have a length that one can measure in the usual way. A famous one is called the "Koch curve" and although it has weird properties it is easy to describe how to make it.

Take an equilateral triangle. At the middle of each side, facing outward, place equilateral triangles one third the size. Now on each side of the resulting figure, place more outward-facing equilateral triangles one third the size of the previous ones. Repeat this process indefinitely, adding smaller and smaller triangles to extend the outer boundary of the figure. The end result is a strange figure indeed, rather like a snowflake in overall appearance. The area it encloses is finite, but the length of its boundary turns out to be $3 \times 4/3 \times 4/3 \times 4/3 \ldots$, which diverges to infinity. Curves like this are known as *pathological curves*. The word "pathological" means diseased, or sick. It is a good name for them.

There is a special term reserved for the boundary dimension of such finite/infinite objects, and it is called the *Hausdorff-Besicovitch* measure. That's a bit of a mouthful. The boundaries of the Mandelbrot set have a fractional Hausdorff-Besicovitch measure, rather than the usual dimension (1) of the boundary of a plane curve, and most people now prefer to use the term coined by Mandelbrot, and speak of *fractal dimension* rather than Hausdorff-Besicovitch dimension. Objects that exhibit such properties, and other such features as scaling, were named *fractals* by Mandelbrot.

Any discussion of chaos has to include the Mandelbrot set, scaling, and fractals, because it offers by far the most *visually* attractive part of the theory. I am not convinced that it is as important as Feigenbaum's universality. However, it is certainly beautiful to look at, highly suggestive of shapes found in Nature and—most important of all—it tends to show up in the study of systems that physicists *are* happy with and impressed by, since they represent the result of solving systems of differential equations.

11.6 Strange attractors. This is all very interesting, but in our discussion so far there is a big missing piece. We have talked of iterated functions, and seen that even very simple cases can exhibit "chaotic" behavior. And we have also remarked that physical systems often exhibit chaotic behavior. However, such systems are usually described in science by *differential equations*, not by iterated functions. We need to show that the iterated functions and the differential equations are close relatives, at some fundamental level, before we can be persuaded that the results we have obtained so far in iterated functions can be used to describe events in the real world.

Let us return to one simple system, the pendulum, and examine it in a little more detail. First let's recognize the difference between an idealized pendulum and one in the real world. In the real world, every pendulum is gradually slowed by friction, until it sits at the bottom of the swing, unmoving. This is a single point, termed an *attractor* for pendulum motion, and it is a *stable* attractor. All pendulums, unless given a periodic kick by a clockwork or electric motor, will settle down to the zero angle/zero speed point. No matter with what value of angle or speed a pendulum is started swinging, it will finish up at the stable attractor. In mathematical terminology, all points of phase space, neighbors or not, will approach each other as time goes on.

A friction-free pendulum, or one that is given a small constant boost each swing, will behave like the idealized one, swinging and swinging, steadily and forever. Points in phase space neither tend to be drawn towards each other, nor repelled from each other.

But suppose that we had a physical system in which points that *began* close together tended to *diverge* from each other. That is the very opposite of the real-world pendulum, and we must first ask if such a system could exist.

It can, as we shall shortly see. It is a case of something that we have already encountered, a strong dependence on initial conditions, since later states of the system differ from each other a great deal, though they began infinitesimally separated. In such a case, the attractor is not a stable attractor, or even a periodic attractor. Instead it is called a *strange attractor*.

This is an inspired piece of naming, comparable with John Archibald Wheeler's introduction of the term "black hole." Even people who have never heard of chaos theory pick up on it. It is also an appropriate name. The paths traced out in phase space in the region of a strange attractor are infinitely complex, bounded in extent, never repeating; chaotic, yet chaotic in some deeply controlled way. If there can be such a thing as controlled chaos, it is seen around strange attractors.

We now address the basic question: Can strange attractors exist mathematically? The simple pendulum cannot possess a strange attractor; so far we have offered no proof that *any* system can exhibit one. However, it can be proved that strange attractors do exist in mathematically specified systems, although a certain minimal complexity is needed in order for a system to possess a strange attractor. We have this situation: Simple equations can exhibit complicated solutions, but for the particular type of complexity represented by the existence of strange attractors, the system of equations can't be *too* simple. To be specific, a system of three or more nonlinear differential equations can possess a strange attractor; less than three equations, or more than three linear equations, cannot. (The mathematical statement of this fact is simpler but far more abstruse: A system can exhibit a strange attractor if at least one Lyapunov exponent is positive.)

If we invert the logic, it is tempting to make another statement: *Any physical system that shows an ultra-sensitive dependence on initial conditions has a strange attractor buried somewhere in its structure.*

This is a plausible but not a proven result. I am tempted to call it the most important unsolved problem of chaos theory. If it turns out to be true, it will have a profound unifying influence on numerous branches of science. Systems whose controlling equations bear no resemblance to each other will share a *structural* resemblance, and there will be the possibility of developing universal techniques that apply to the solution of complicated problems in a host of different areas. One thing in common with every problem that we have been discussing is *nonlinearity*. Nonlinear systems are notoriously difficult to solve, and seem to defy intuition. Few general techniques exist today for tackling nonlinear problems, and some new insight is desperately needed.

If chaos theory can provide that insight, it will have moved from being a baffling grab-bag of half results, interesting conjectures, and faintly seen relationships, to become a real "new science." We are not there yet. But if we can go that far, then our old common sense gut instinct, that told us simple equations must have simple solutions, will have proved no more reliable than our ancestors' common sense instinctive knowledge that told them the Earth was flat. And the long-term implications of that new thought pattern may be just as revolutionary to science.

Today, we are in that ideal time for writers, where what can be speculated in chaos theory far exceeds what is known. I still consider the ultimate importance of chaos theory as not proven, but it has certainly caused a change of outlook. Today you hear weather forecasters referring to the "butterfly effect," in which a butterfly flapping its wings in the East Indies causes a hurricane in the Caribbean—a powerful illustration of sensitive dependence on initial conditions.

Science fiction writers long ago explored the idea of the sensitive dependence on initial conditions in time travel stories. In "A Sound of Thunder" (Bradbury, 1952), a tiny change in the past produces a very different future.

Is time really like that? Or would small changes made in the past tend to damp out over time, to produce the same present? Stating this another way, if we were to rerun the history of the planet, would the same life-forms emerge, or a totally different set? This is a hot topic at the moment in evolutionary biology.

CHAPTER 12
Future War

Pessimists, gloomy about the future, point out that the only continuous progress in human history seems to be in methods of warfare. Optimists point out that humans are still here, most of us living far better than our ancestors ever dreamed. Pessimists reply, ah, but wait until the next war.

Einstein, asked about the weapons of the Third World War, said that he did not know what they would be; but that the Fourth would be fought with sticks and stones.

Einstein died in 1955. Were he alive today, I think he would be both gratified and horrified. Gratified, because the all-consuming fear of the 1950s was of large-scale nuclear war. Not only have we escaped that, but after the end of the Second World War we have avoided the use of nuclear weapons in combat. But Einstein would surely be dismayed at the continuing conflicts, all around the world, and the increased vulnerability of cities and civilians to terrorist acts.

Dismayed, too, at the potential of today's science for the creation of new weapons. Weapons drive, and are driven by, technological advances. If the scale of war remains small, weapons are likely to become more tricky, more deadly—and more personal.

Anyone who reads my stories may suspect that war and military affairs are not among my main interests.

That is true. However, military science fiction is a big
component of the field, and many people read little
else. Regard this chapter, then, not as a compendium
of military knowledge, or a source book on the writing of military science fiction. Think of it as a discussion of a few story ideas with military potential that
seem to have been overlooked.

12.1 The Invisible Man. The best weapon of all is one
which the adversary never realizes has been used.

Wouldn't it be wonderful if you could put on your
tarnhelm and, like the old Norse heroes, become invisible
to your enemies? You could sneak through lines of
defense, sit in on private strategy meetings, steal battle
plans, and even kill selected people. I have no doubt
that our leaders in the Pentagon and CIA Headquarters,
gnashing their teeth over their inability to get to Saddam
Hussein, would have given a fair number of those teeth
for a good cloak of invisibility.

Can it be done?

H.G. Wells took a shot at the problem in his novel
The Invisible Man (Wells, 1897). His solution, a drug
to make every part of the human body of the same
refractive index as air, possesses a number of difficulties
that I feel sure Wells knew about. Let us put aside
the improbability of the drug itself, and examine other
effects.

First, if no part of your body absorbs light, that
includes the eyes. Light will simply pass through them.
But if your eyes do not absorb light, you will be blind.
Seeing involves the absorption of photons by your retinas.

Then there is the food that you eat. What happens
to it while it is being digested? It would be visible in
your alimentary canal, slowly fading away like the
Cheshire Cat as it went from esophagus to stomach and
to intestines.

I think Wells could have done better with a little
thought; and we can do a lot better, because we have
technology unknown in his day.

Consider how Nature tackles the problem of invisibility.
The answer is not drugs, but deception. Animals do their

best to be invisible to their prey or their predators. But they don't do it by fiddling around with their own internal optical properties. They are invisible if they look exactly like their background. The chameleon has the right idea, but it's hardware-limited. It can only make modest color and pattern adjustments. Humans disguise their presence with camouflage, but that, too, is a static and simple-minded solution.

What we need is a whole-body suit. The rear part of the suit is covered with tiny imaging sensors, admitting light through apertures no bigger than pinholes. Their outputs feed charge-coupled devices, which pass an electronic version of the scene behind you, in full detail, to the suit's central processing unit (cpu). The front of the suit contains millions of tiny crystal displays. The cpu sends to each of these displays an appropriate signal, assigning a particular color and intensity. Seen from in front, the suit now mimics its own background (the scene behind it) perfectly.

So far this is straightforward, comfortably within today's technology. The difficulty comes because the suit has to reproduce, when viewed from any angle, the background as seen from that angle. Someone behind you, for example, has to see an exact match to the scene in front of you. To get the right effect from all angles, you have to use holographic methods and generate multi-angle reflectances. The computing power to do all this is considerable—far more than anything in today's personal computers. However, we have seen where that technology is going. Twenty or thirty more years, and the computing capacity we need will probably fit into your wristwatch.

The problem of vision, never addressed by Wells, is also easy enough. The signal received from in front of the suit is pipelined to goggles contained within the suit's helmet. We would anticipate that a suit like this would work only with uniform, low-level illumination and relatively uniform backgrounds.

Could we build one, today? I don't know, but if we could and we tried to sell it, I bet that its use would be banned in no time.

12.2 Death rays. The death ray was introduced to science fiction in its early days. H.G. Wells was responsible for this one, too. The beam of intense light, flashing forth to set fire to everything in its path, is something we all remember from *The War of the Worlds* (Wells, 1898). Wells called it a heat ray, and he described it in the language of weapons: "this invisible, inevitable sword of heat."

For the next half century, every respectable scientist knew that such a ray was impossible. Science fiction writers of the 1930s, however, continued to use it freely. And hindsight proves that they were right to so do. The three scientific papers that permit the death ray—we now call such a thing a laser—to exist had been published in 1916 and 1917, ironically at the height of the "war to end wars." The papers were by Einstein, and they established balances among the rates at which electrons orbiting an atom can move to higher or lower energy levels.

This requires a little explanation. In Chapter 5, we noted that electrons around an atomic nucleus sit in "shells," and that an element's freedom to react chemically depends on whether it has a filled or a partially empty outer electron shell.

In addition to the locations where electrons are normally situated, there are other possible sites where an electron can reside temporarily. An electron can be boosted to occupy such a site, provided that it is supplied with energy in the form of radiation. If an electron is in such a higher-energy position, it is said to be in an *excited state*. An electron with no extra energy is said to be in its *ground state*.

Left to itself, an electron in an excited state will drop back to its ground state, emitting radiation as it does so. This is known as *spontaneous emission*. The return to the ground state normally happens quickly, but that is not always the case. Sometimes an electron can be at an excited energy level where other physical parameters, such as orbital angular momentum or spin, are incompatible with a straightforward return to the ground state. Such a transition is known as a *forbidden transition*, and the effect is to make the electron remain longer in the excited state.

Even a forbidden transition normally takes place in a fraction of a second. The phenomenon that most of us have seen, called *phosphorescence*, in which a material continues to glow after it has been removed from sunlight, is a more complicated process. In phosphorescence, the electron usually becomes "trapped" in a dislocation in a crystal lattice structure. Only after it leaves that trap (which may take minutes or even hours) can it finally undergo spontaneous emission to the ground state.

An electron can be induced to make a forbidden transition from a higher energy state to a ground state, by providing to it radiation of a suitable wavelength. This is known as induced or *stimulated emission,* because the electron as it drops back to the ground state gives out radiation of an energy appropriate to that state change.

We now have all the ingredients for our basic death ray. We pump energy into a material, raising large numbers of electrons to excited states. They will fall back by spontaneous emission to a lower energy. However, if we have picked the right material many of them do not go at once to the ground state. Instead, they drop to another excited state from which the transition to the ground state is forbidden. There they stay, increasing in numbers, until at last we supply radiation of the right wavelength to induce a fall to the ground state. They do this in large numbers, producing a huge pulse of released energy in the form of light. The emitted light is monochromatic (of a single, precise wavelength) and coherent (all of the same phase).

This is today's laser—light amplification by stimulated emission of radiation. The first one was built in 1960, and they are used now for everything from data transmission to eye surgery.

This may seem somewhat disappointing for something billed as a "death ray." However, lasers are certain candidates for future wars. The first lasers were of low power, but that has changed. Great power (kilowatts and more) can be delivered into very small areas. A laser beam will burn almost instantly through any known material, and by 1968 it had already been used to initiate thermonuclear reactions. Perhaps even more relevant for the purposes of war, a laser beam can be made very narrow, with little spread over large distances. Since the power is delivered to the target at the speed of light, high-energy lasers are good in either offense or defense and have been proposed as the most effective form of protective shield against missile attack. They also, because they remain as a tight beam over great distances, have been suggested as the best way of launching spacecraft, or of sending power to them anywhere in the solar system.

The first lasers employed electrons in the outer atomic shells of the atom, and the radiation they produced was normally in the visible or near-infrared wavelength regions of the spectrum. However, there is no reason in principle why an electron in the inner electron shells should not go through the same processes of energy absorption, spontaneous emission to a forbidden state, and final stimulated emission. Because the inner electron shells are more tightly bound, the energy released on the final return to ground state is higher, and the wavelength of the radiation produced will be shorter. The result is an X-ray laser: invisible in its output, and considerably more deadly.

12.3 The ultimate personal weapon. War isn't what it was. In ancient times, one rational and economical way of deciding the outcome of a battle was through the use of champions.

You select the best fighter in your army. I do the same in mine. We let those two fight it out, while the rest of us stand around, watch, and cheer on our guy. The individual serves as a surrogate for the whole army. If he wins, we all win; if he loses, we admit defeat.

I don't think this was ever a common method—suppose I have ten times as many soldiers as you, but you have one huge chap twice the size of any of my people? Do you think I am going to risk losing the whole war with a one-on-one fight? Even if it worked against Goliath, I don't want anything to do with it.

Individual combat, by chosen champions, will certainly not work today. For one thing, our weapons make personal strength in combat rather irrelevant. But the combat of champions makes a point that is as valid now as it ever was: in war, as in all other human activities, individuals make a difference. Some wars arise because of the ambitions of a single human.

Such a person is usually well-protected, and capture is difficult. Killing is easier. A 20-megaton hydrogen bomb in downtown Baghdad would almost certainly have taken care of Saddam Hussein. But how many hundreds of thousands of innocents would have been killed along with him?

The sledgehammer-on-the-ant solution is no solution at all. Too many would die. But suppose we could, neatly and cleanly, dispose of the major troublemaker.

It was tried with Adolf Hitler, and failed. A bomb carried into a conference exploded as planned, but he was shielded from the blast by the leg of a table. It was tried by the CIA with Fidel Castro, and produced a variety of failures that read like a catalog of ineptitude. (Poisoned cigars, no less. Shades of Snow White.)

But does it have to fail? Or can we suggest ways to guarantee the death of a single, chosen individual?

Let us go back to basic biology. In what way is El Supremo, busy causing so much trouble, different from every other person on Earth?

Forget photographs, forget fingerprints and retinal patterns. More reliable than either, and increasingly recognized as such in court cases, is the uniqueness of an individual's DNA. Unless you happen to have an identical twin, your DNA is yours and yours alone.

It should be a lot easier to obtain a sample of El Supremo's DNA than it would be getting close to the man himself. An old hat, a sock, or a dirty shirt will contain little flakes of skin, a razor may have a tiny drop of dried blood. Remember, a single cell should be enough. The entire genome is in every nucleus.

Inside El Supremo's body, just as in your or my body, there are defense cells known as *T cells*. Their job is to mop up viruses that have invaded the body. This happens all the time, because viruses are small and light enough to be airborne. We take in viruses with every breath, and our T cells destroy them. If you have few T cells, your body will lose its resistance to outside infection, which is exactly what happens to people with AIDS.

However, the T cells can't go around destroying everything in sight. They have to be able to distinguish foreign matter, which is not supposed to be there, from the cells of your own body. If something looks like you, in the right kind of way, your T cells will not touch it.

What kind of way is that? Well, among other things our DNA contains a sequence that codes for the production of a molecule called a *major histocompatibility complex* (MHC). The MHC in your body, like your DNA, is unique and specific to you. The MHC, which is safe from T cell attack because it is recognized as part of you, can carry other things to a T cell. Those other things will then also be judged as part of you, and left in peace; or as alien to you, and destroyed.

Now we are ready to go to work. Recall, from Chapter 6, that a virus is little more than a package of DNA wrapped in a coat of protein. We will take a virus that contains the DNA of a lethal disease. There are plenty of those. However, we will give to that virus a protein coat matching the MHC profile of a specific person— say, El Supremo.

We reproduce the virus in large quantities, and release it in the city where El Supremo makes his home. People breathe it in, and it enters their bodies. If it begins to reproduce in any quantity, the T cells recognize its MHC coat as alien to the body (or rather, to that body's genetic code). They destroy the virus. The numbers of the virus remain in check. People breathe some out, which are in turn breathed in by other people.

This goes on—until, weeks or months later, a sample of the virus is breathed in by El Supremo.

At this point, everything changes. The virus is recognized by the T cells as part of El Supremo, because its protein coat looks like the right MHC molecules. The virus can continue the process of cell invasion and reproduction, and the T cells will leave it alone. A few days later, the tragic death of El Supremo is reported—struck down in his prime by a fatal viral infection.

There is no suspicion of foul play. No one else becomes sick, or is in any danger. We employed a weapon that could harm no one in the world but El Supremo.

Can we do all this, today? Not quite. But when the human genome mapping project is completed, a few years from now, we should have the genetic maps and the rapid DNA sequencers to do the job. War will become a more personal tool than it has ever been in the past.

And war will take on a more generalized meaning. The use of the basic idea presented here is described in the novel *Oaths and Miracles* (Kress, 1996)—with organized crime controlling the biological weapon.

12.4 Cyborgs. A cyborg is a rather ill-defined amalgamation of human and mechanical components. Most of us today are cyborgs. I am wearing glasses, and I have one gold tooth and another that has been capped with some unknown (to me) white material. I have friends with artificial hips and knees, a couple of others wear pacemakers. Most teenagers I meet have braces on their teeth.

By science fiction standards this is pretty pathetic stuff. When we say "cyborg," we expect at the very least something like the Six Million Dollar Man, capable of feats of strength and speed beyond the human. If it is to be war—the enhanced warrior is a common feature of science fiction—then we demand the super-soldier, augmented and improved with built-in superhuman fighting equipment, like the speeded-up Gully Foyle in *The Stars My Destination* (Bester, 1956).

Is such a creation possible?

It may be, but it is not so clear that it is desirable. If you use a cyborg in a story you will find yourself constrained by the laws of both physics and biology. The human body, considered as raw material for war, has some severe drawbacks. If we have a choice of a human-machine combination, rather than pure human or pure machine, which human features would we keep?

TABLE 12.1 (p. 300) offers a comparison of human

and machine properties and limitations. We assume that there will be significant change in the next forty or fifty years in what computers and robots can do, but little change in human capabilities.

We see that our big advantages lie in our versatility, built-in repair capability, and self-reproducing feature. Ideally, the cyborg will have the power to repair or reproduce itself. In the near future, the human part can do the repairs to the machine, but what about production of additional copies? A machine that repairs and reproduces itself from available raw materials is certainly a long-term possibility. But what then is the role of the human, as robotic reasoning powers increase?

There is also the problem of mismatch, particularly on variables like physical strength. We can, if we choose, give our cyborg a "bionic arm," able to lift tons. But if the rest of the body is still flesh and bone, use of the bionic arm will produce intolerable stresses. The same problem will arise if we speed up the human reaction time too far. Our natural reaction speed is close to the limit of what our bodies can stand. The pulled muscles of Olympic sprinters attest to this.

We may benefit by giving our cyborg enhanced sensory powers—it might be useful to communicate via bat-squeak signals, or see thermal infrared radiation—but we are unlikely to gain from superior strength unless it is accompanied by the installation of superhuman muscles and bones. One way to do this is via an exoskeleton to take the added stresses. However, if we are going outside the body with our improvements, we may as well go all the way and put ourselves inside a car or tank or spaceship. That has another great advantage. If the machine goes wrong, we can abandon it and get ourselves another one.

My final conclusion: half a century from now, a cyborg will be regarded as an inferior form of robot. As a warrior, it will lose every fight to its totally inorganic foes.

12.5 Cleaning up after nuclear war. Since 1945, the horrors of nuclear war have been graphically depicted in books and movies. After a while, recitation of biological and physical effects becomes numbing. It is hard for a rational mind to distinguish the effect of a 10-megaton TNT-equivalent hydrogen bomb from a 50-megaton TNT-equivalent bomb, because they are both so intolerable.

Unfortunately, that does not mean they will never be used. Suppose that, despite all our efforts, a nuclear bomb (large or small) is exploded. One of the things that makes such weapons worse than their chemical equivalents is the longevity of their aftereffects. Some radioactive by-products have very long halflives. After the bombs have exploded, perhaps even after the causes of the war are forgotten, radioactive debris will remain.

Must this be so? Must lands remain blighted for a hundred or a thousand years? Actually, all the evidence so far suggests no such thing. Life appears to be far too resourceful and adaptable to be discouraged by high ambient radioactivity. Plants and animals are thriving on the Pacific atolls where hydrogen bombs exploded forty years ago, and new growth appears in the very shadow of the Chernobyl reactor building. However, let us assume that radioactive waste is at least a nuisance. Can we see ways in which our descendants might cleanse radioactive ruins faster than Nature seems to permit?

There is a way, at least in principle, with a "nuclear laser" that we don't yet know how to build.

To understand how the device might work, consider the nature of radioactivity. The nucleus of a radioactive atom is unstable. It is, in energy terms, not in its ground state. After a shorter or longer time period, the nucleus will emit a particle which allows it to proceed to a state of lower energy. That state may actually be a different element. For example, if a nucleus emits an electron it will change to become the element next higher on the atomic table, since it now, in effect, has one more proton than before. If it emits an alpha particle (the nucleus of the helium atom) it will become the element two lower in the atomic table.

The state of the nucleus after emission of a particle may be a "ground state," which means that the nucleus is now stable; or it may be a transitional state, in which case after a shorter or longer period another radioactive decay will take place, moving the nucleus to a lower energy level and again perhaps causing it to become a different element.

The whole process of radioactive decay can be viewed as a nucleus moving to states of successively lower and lower energy, until it finally rests in the (stable) ground state. Although we know on average how long it takes a nucleus to go from one state to another—this is what the halflife measures, the time for half the nuclei in a large sample to make the change of state—there is no way to predict when any particular nucleus will undergo radioactive decay.

As an example of the sequence of changes, consider plutonium. Element 94, it is not found naturally on Earth. It is presumably created, like other heavy elements, in supernova explosions, but it is radioactive and its most stable form, $_{94}Pu^{244}$, has a halflife of less than a hundred million years. Here is the complete decay sequence, with the halflife indicated for each step, for a more short-lived form of plutonium:

$_{94}Pu^{241}$ (13.2 years) to $_{95}Am^{241}$ (458 years) to $_{93}Np^{237}$ (2 million years) to $_{91}Pa^{233}$ (27 days) to $_{92}U^{233}$ (160,000 years) to $_{90}Th^{229}$ (7,340 years) to $_{88}Ra^{225}$ (14.8 days) to $_{89}Ac^{225}$ (10 days) to $_{87}Fr^{221}$ (4.8 minutes) to $_{85}At^{217}$ (0.032 seconds) to $_{83}Bi^{213}$ (47 minutes) to $_{84}Po^{213}$ (4.2 microseconds) to $_{82}Pb^{209}$ (3.3 hours) to $_{83}Bi^{209}$.

Look like gibberish? Sorry. Nuclear physicists and chemists deal with sequences like this every day.

Bismuth, $_{83}Bi^{209}$, is stable. The intermediate elements referred to in this radioactive decay chain are lead (symbol Pb, element 82), polonium (Po, element 84), astatine (At, element 85), francium (Fr, element 87), radium (Ra, element 88), actinium (Ac, element 89), thorium (Th, element 90), protoactinium (Pa, element 91), uranium (U, element 92), neptunium (Np, element 93), and americium (Am, element 95).

What can we say about this rather messy chain of decay products? Well, almost all the time for the whole decay process comes with transitions that take 13.2 years, 458 years, 2 million years, 160,000 years, and 7,340 years.

If we could somehow get rid of those, our dangerous plutonium $_{94}Pu^{241}$ would turn to stable bismuth $_{83}Bi^{209}$ in a few months.

Is there any way to draw the fangs of radioactivity, by speeding up the long transitions?

We return to the similarity between the atom and the nucleus, discussed in Chapter 5. There, we noted that the protons and neutrons of the nucleus, like the electrons of the atom, can be thought of as fitting into shells. There are excited states of the nucleus, as there are excited states of the atom, and for protons and neutrons such excited states of the nucleus are termed *resonances*. The whole phenomenon of radioactivity can be thought of as a nucleus, descending to its ground state by a series of transitions. "Forbidden" nuclear transitions are marked by very long decay times.

What we need—but do not yet have, with the science of today—is a mechanism for *stimulated nuclear emission*. We need a nuclear laser, in which the application of radiation (very short wavelength) triggers a stimulated nuclear transition and accomplishes in minutes what normally takes thousands or millions of years. This would call for careful control, because all the energy provided by the transitions could be given off in a very short time. The process would have to be timed carefully, drawing off energy at a tolerable rate.

However, there is another good side to this. We have found a potentially useful supply of free energy. In one possible future, maybe our despised dumps of radioactive wastes will be as sought after as today's oil fields.

TABLE 12.1
Comparison of human and machine performance and tolerances.

Property	Human	Machine
Gravity field/accn.	0.25–2 gee*	0–50,000 gee
Hearing range	20–20,000 cps	0–10^6 cps
Vision detail	Resolves to one minute of arc	Can easily resolve to one second of arc
Vision wavelength range	red to violet	X-rays to long wave radio
Air pressure	0.3–1.5 atms.	0–100,000 atms.
Necessary support	Air, food, water	Power supply
Operating rate	Almost fixed	Variable; includes zero rate (off)
Speed of thought	1,000 cycles/sec	> 10^{12} cycles/sec
Mass	15–500 kg**	Any value
Radiation tolerance	Poor	Easily hardened to radiation
Mean time to failure	< 100 years	Variable; no reason why it should not be > 1,000 years
Strength	Poor	To limits of material science
Versatility	Excellent	Very limited
Repairs	Automatic	Needs external maintenance
Production/ Reproduction	Easy***	Needs factory

* A human can survive in free-fall, but physical deterioration, including bone loss, usually sets in after months in a low gravity field.

** There have been humans who weigh as little as 15 kg (33 pounds) and as much as 500 kg (1,100 pounds). I'm not sure either would be my choice for a warrior.

*** Perhaps too easy.

CHAPTER 13
Beyond Science

13.1 Scientific heresies. The intent of this book is to define the boundaries of science, in order that we can then consider going beyond them. This chapter, however, is a little different. Here we will consider "scientific heresies," ideas that are already beyond the boundaries of accepted science.

I can make an argument that the place where any self-respecting scientist ought to work is in precisely the fringe areas where the uncertainty is greatest and our knowledge is least. Scientific heresies often appear when science has reached some kind of sticking point, but most scientists are reluctant to admit it. These are the areas in which the great leaps forward are likely to be made. In practice, most young scientists cannot afford to take the chance. The probability of failure is too great. Einstein said that the reason he could spend the last quarter-century of his life in a search (unsuccessful) for a unified field theory was that he could afford to; his reputation was already secure. A younger scientist takes a great risk by choosing to work in any field of scientific heresy.

That does not mean the heretical ideas are necessarily wrong.

I give the previous sentence special emphasis, because when I wrote on this subject a couple of years ago I received outraged letters defending people's

favorite theories. How dare I say that such and such
an idea was heresy?

In each case, I had to write back and explain: A
"scientific heresy" is a theory which runs contrary to
the accepted scientific wisdom of a particular time. It
is no more than that. It may eventually prove to be an
improved description of Nature, so that it later (and often
after much argument) becomes part of the standard world
view; or it may prove to be misguided, and join the large
group of discredited crank theories.

Suppose a scientific idea is heresy. Is it wrong? I don't
know. What I do know is that if most scientists, today,
don't accept it, that is enough to make it a scientific
heresy.

As a first example, I offer something that fifteen years
ago was certainly heresy. Now it seems on the way to
becoming scientific dogma and uncritically accepted fact.

13.2 Dinosaur doom. As proved by the huge success
of *Jurassic Park*, people love dinosaurs. It is a long-lasting
love affair. Fifty years ago, one of the most memorable
segments of Walt Disney's movie *Fantasia* involved these
animals. First we see dinosaurs feeding, raising their
young, and hunting in a humid swamp world of frequent
rains. Then we see a climate change, to a barren desert
where the dust-clouded sun shines constantly. The water
has all vanished, and nothing but dry bones and dead
trees remain.

This was an artistic portrayal of one of Nature's greatest
mysteries: Why did the dinosaurs, the dominant land
animal for one hundred and fifty million years, disappear?
Moreover, why did they vanish so abruptly and so
completely?

Fantasia's answer, extreme climate change, introduced
more questions than it answered. Clearly, our planet's
lands did not all change to a dry and barren desert
devoid of plants. Had that happened, the conquering of
the land by plant forms would have had to start over
from scratch, and there is no sign of such a thing in
the fossil record. Moreover, since the dinosaurs were
widespread, with their remains found everywhere from

America to China, any climate change would have to be not only severe, but ubiquitous. More recent suggestions, that dinosaurs were warmblooded creatures, make their worldwide extinction by climate change even more unlikely. Finally, we have to ask why the disappearance of the dinosaurs happened so quickly.

Even fifty years ago, when *Fantasia* was produced, it was possible to object to the picture painted in the Disney movie. On the other hand, no one had a better answer. Climate change, perhaps accompanying a period of high volcanic activity, seemed like the best available explanation for what is referred to in scientific circles as "the K/T extinction." The K/T boundary is the time when the geological time period known as the Cretaceous Period ended and the Tertiary Period began. Why K/T, rather than C/T? Because geologists refer to geological strata using a single letter, and C is already taken for the Carboniferous Period (which ended about 140 million years before the Cretaceous Period began). At the same time as the dinosaurs vanished from the Earth, so too did all the large flying reptiles and the largest marine reptiles. It is logical to assume that the cause of all three disappearances was the same, on land, air, and sea.

A new possibility appeared in 1978, when a son brought home a rock sample and showed it to his father. The son was Walter Alvarez, a geologist and geophysicist and a professor at the University of California at Berkeley. The father was the late Luis Alvarez, a physicist who had won the 1968 Nobel Prize for his contributions to elementary particle physics. The rock had been dug by Walter Alvarez himself from a gorge in the Italian Apennines. Its lower part was white limestone, its middle a half-inch layer of hardened clay, its upper part red limestone. The boundaries between the three sections were quite clearly defined. Walter Alvarez knew that a similar clay layer was to be found all around the world. The clay had been laid down on the ocean floor sixty-five million years earlier—right about the time that paleontologists believed that the dinosaurs became extinct.

Luis Alvarez thought it might be possible to determine how long the clay layer had taken to form by a technique involving induced radioactivity, and he and his son looked in the rock sample's clay layer for the presence of a suitable substance. Sure enough, they and their co-workers found iridium—but they discovered three hundred times as high a concentration of that element in the clay as in the limestone layers above and below it. Since iridium is rare in the Earth's crust and much more common in meteorites that fall to the Earth, it seemed possible that the clay layer might be tied in with some major extraterrestrial event.

After Luis Alvarez ruled out a couple of other candidate ideas, including a nearby supernova explosion, a colleague of the Alvarez's, Chris McKee, suggested that an asteroid, maybe ten kilometers across, might have hit the Earth. That size of asteroid was consistent with the observed iridium content of the clay layer. The energy release of such an impact would equal that of a hundred million medium-sized hydrogen bombs. The effect of the impact would be the disintegration of the asteroid, which along with many cubic kilometers of the Earth's crust would have been thrown high into the atmosphere. The dust would have stayed there for six months or so, halting photosynthesis in plants, preventing plant growth, and thereby starving to death all the larger animals. Upon the dust's return to Earth it would create the thin layer of clay seen between the red and white limestones. The same phenomenon of atmospheric dust had been observed on a much smaller scale following the huge volcanic explosion of Krakatoa in 1883. It is also the mechanism behind the idea of "nuclear winter," a palling of the whole Earth with atmospheric dust which some scientists worry could be one consequence of a large-scale nuclear war.

Twenty years ago, however, the idea of nuclear winter had not been taken seriously. Thus the theory presented by Luis and Walter Alvarez for the vanishing of the dinosaurs was pure scientific heresy. It was widely criticized by paleontologists, and even ridiculed. Today it is widely accepted as the most plausible extinction mechanism, and the same idea has been used to examine other major disappearances of many life forms from Earth. The greatest of these, known as the Permian extinction, occurred about 230 million years ago, when nine-tenths of all Earth's species vanished. The search for asteroid evidence has been less persuasive in this case.

Also unresolved in the case of the K/T extinction is the question of where the incoming destroyer hit the Earth. The most popular theory at the moment is that it struck in what is now the Gulf of Mexico, but that is not fully proved.

Many other details of the asteroid impact theory remain to be defined. However, those who do not believe the idea at all must face one inevitable and awkward question: If it was not the impact of a huge asteroid that suddenly and swiftly killed off all the dinosaurs, then what was it?

If you think it would be a nice idea to write a story in which a large asteroid descends on Earth today and causes all sorts of problems, be warned. The book has already been written, several times. I'll mention just two examples: *Lucifer's Hammer* (Niven and Pournelle, 1977), and *Shiva Descending* (Benford and Rotsler, 1980). If you want to read about a large object hitting the Moon, and what that can do to Earth, read Jack McDevitt's splendid *Moonfall* (1998).

13.3 Gaia: the Whole Earth Mother. This, too, borders on scientific respectability, though scientists as well-known as Stephen Jay Gould and Richard Dawkins have dismissed it as pseudoscience.

It began in the late 1970s, when James Lovelock published a controversial book, *Gaia: A New Look at Life on Earth* (Lovelock, 1979). In it he set forth his idea, long gestating, that the whole of Earth's biosphere should be thought of as a single, giant, self-regulating organism, which keeps the general global environment close to constant and in a state appropriate to sustain life. In Lovelock's own words, *Gaia* is "the model, in which the Earth's living matter, air, oceans and land surface form a complex system which can be seen as a single organism and which has the capacity to keep our planet a fit place for life."

Lovelock says that the notion is an old one, dating

back at least to a lecture by James Hutton delivered in
1785. However, the modern incarnation of that idea is
all Lovelock's, although the name *Gaia* as a descriptor
for such an interdependent global entity was provided
by the late William Golding (a Nobel laureate for
literature, Lovelock's neighbor in England, and author
of the classic *Lord of the Flies*).

Something like Gaia seems to be needed from the
following simple physical argument: Life has existed on
Earth for about three and a half billion years. In that
time, the sun's energy output has increased by at least
thirty percent. If Earth's temperature simply responded
directly to the Sun's output, based on today's global
situation we would expect that two billion years ago the
whole Earth would have been frozen over. Conversely,
if Earth was habitable then it should today be too hot
to support life.

But in fact, the response of Earth's biosphere to
temperature changes is complex, apparently adapting to
minimize the effects of change. For example, as the
amount of solar energy delivered to Earth increases,
the rate of transpiration of plants increases, so the
amount of atmospheric water vapor goes up. That
means more clouds—and clouds reflect sunlight, and
shield the surface, which tends to bring surface temperatures down. In addition, increased amounts of
vegetation reduce the amount of carbon dioxide in the
air, and that in turn reduces the greenhouse effect by
which solar radiation is trapped within the atmosphere.
Again, the surface temperature goes down. There are
many other processes, involving other atmospheric gases,
and the net effect is to hold the *status quo* for the
benefit of living organisms. According to Lovelock, it
is more than a matter of convenience. Only the presence of life has enabled Earth to remain habitable. If
life had not appeared on this planet when it did, over
three billion years ago, then by this time the surface
of Earth would be beyond the range of temperatures
at which life could exist.

Why, then, does the *Gaia* idea qualify as a scientific
heresy? It sounds eminently reasonable, and something

like it seems necessary to explain the long continuity of life on the planet.

Part of the problem is that at first thought it seems as though the whole Earth must be engaged in some sort of activist role. Many readers have assumed that *intention* is a necessary part of the *Gaia* idea, that the biosphere itself somehow knows what it is doing, and acts deliberately to preserve life. A number of non-scientific writers have embraced this "Earth as Ur-mother" thought in a way and with an enthusiasm that Lovelock neither intended nor agrees with. At the other extreme, two biologists, Doolittle and Dawkins, have offered the rational scientific criticism that the *Gaia* idea seems to call for global altruism, i.e. some organisms must be sacrificing themselves for the general good. That runs contrary to everything we believe to be true about genetics and the process of evolution.

Lovelock seemed at first to encourage such a viewpoint, when he wrote, "But if Gaia does exist, then we may find ourselves and all other living things to be parts and partners of a vast being who in her entirety has the power to maintain our planet as a fit and comfortable habitat for life." There is more than a suggestion here of a being which acts by design. However, Lovelock has later shown through simplified models that neither global intention nor global altruism is needed. The standard theory of evolution, in which each species responds in such a way as to assure its own survival and increase its own numbers, is sufficient to create a self-stabilizing total system.

Today the *Gaia* hypothesis, that the whole Earth biosphere forms a single, self-regulating organism, is still outside the scientific mainstream. However, over the past fifteen years it has gained some formidable supporters, notably the biologist Lynn Margulis, who has championed *Gaia* more actively than Lovelock ever did. The theory also provides a useful predictive framework for studying the way in which different parts of the biosphere interact, and particular chemicals propagate among them. Nonetheless, if it is not today outright heresy, to many scientists *Gaia* remains close to it.

Lovelock ironically comments that we may have come " . . . the full circle from Galileo's famous struggle with the theological establishment. It is the scientific establishment that now forbids heresy. I had a faint hope that *Gaia* might be denounced from the pulpit; instead I was asked to deliver a sermon on *Gaia* at the Cathedral of St. John the Divine in New York."

The *Gaia* concept sometimes permits Lovelock to take an unusually detached attitude to other global events. Some years ago I was driving him from suburban Maryland to the Museum of Natural History in Washington, D.C. On the way we somehow got onto the subject of all-out nuclear war. Lovelock surprised me very much by remarking that it would have very little effect. I said, "But it could kill off every human!"

He replied, "Well, yes, it might do that; but I was thinking of the effects on the general biosphere."

I leave the subject of *Gaia* with this story idea: suppose that the biosphere did know what it was doing, and acted deliberately to preserve life. How do you think it would deal with humans?

13.4 Dr. Pauling and Vitamin C. When sea voyagers of the fifteenth and sixteenth centuries began to undertake long journeys out of sight of land, and later when Arctic explorers were spending long winters locked in the ice, they found themselves afflicted by a strange and unpleasant disease. Joints ached, gums blackened, teeth became loose and fell out, and bodies showed dark, bruise-like patches. Eventually the sufferers died after a long and painful illness. No one was immune, and as the trip went on more and more people were affected. Thus Vasco da Gama, sailing round the Cape of Good Hope in 1498, lost a hundred of his hundred and sixty crew members. Travelers gave the disease a name, *scurvy*, but they had no idea what caused it.

After many years of trial and error, sea captains and physicians learned that scurvy could be held at bay by including regular fresh fruit and vegetables in the diet. In 1753, the Scottish physician James Lind showed that the same beneficial effect could be produced by the use of concentrated orange and lemon juice. However, no one knew quite what these dietary additives were doing. That understanding had to wait for almost two more centuries, until 1932, when a substance called *Vitamin C*, or *ascorbic acid*, was isolated.

Vitamins are part of our necessary diet, but unlike proteins, carbohydrates, or fats, they are needed only in minute quantities. A daily intake of one thousandth of an ounce of Vitamin C is enough to keep us free from scurvy. Most animals can manufacture for themselves all the Vitamin C that they need; just a few species— humans, monkeys, and guinea pigs—rely on their food to provide it (humans, monkeys, and *guinea pigs*?! A story here, perhaps). Certain foods, such as broccoli and black currants, are especially rich in this vitamin, but almost all fresh fruit and vegetables contain enough to supply human needs. Without it in our diet, however, people sicken and die. Fortunately, Vitamin C is a simple molecule, and by 1933 chemists had learned how to produce it synthetically. It can be made in large quantities and at low cost. No one today needs to suffer from scurvy.

That might seem to be the end of the story of Vitamin C, except that in 1970, the scientist Linus Pauling came forward with an extraordinary claim. In his book *Vitamin C and the Common Cold* (Pauling, 1970), Pauling stated that large doses of Vitamin C, thirty to a hundred times the normal daily requirement, would help to ward off the common cold, or would reduce the time needed for a sufferer to recover.

Most people coming forward with such a notion would have been brushed aside by the medical profession as either a harmless crank, or some charlatan peddling his own patent nostrum or clinic.

There was just one problem. Linus Pauling was a recognized scientific genius. During the 1930s he had, almost single-handed, used quantum theory to explain how atoms bond together to form molecules. For this work he received the 1954 Nobel Prize for Chemistry. Rather than resting on his laurels, he had then gone on to study the most important molecules of biochemistry, in particular hemoglobin and DNA, and was the first person to propose a form of helical structure for DNA.

James Watson and Francis Crick, whom we met earlier in Chapter 6, elucidated the structure of DNA. What did they worry about as they worked? As Watson said in his book, *The Double Helix* (Watson, 1968), they knew that "the prodigious mind" of Linus Pauling was working on the problem at the California Institute of Technology. In the early spring of 1953 they believed that he would discover the correct form of the molecule within a few weeks if they failed to do so. With a little change in timing, or with better experimental data, Linus Pauling might well have won or shared the 1962 Nobel Prize that went to Crick, Watson, and Maurice Wilkins.

However, Pauling had no reason to feel too disappointed in that year. For he was in fact awarded a 1962 Nobel Prize—for Peace, acknowledging his work toward the treaty banning the atmospheric testing of nuclear weapons.

Faced with a two-time Nobel Laureate who was close to being a three-time Laureate, a man still intellectually vigorous at age 69, the medical profession could not in 1970 dismiss Pauling's claims out of hand. Instead they investigated them, performing their own controlled experiments of the use of Vitamin C to treat the common cold. Their results were negative, or at best inconclusive.

That should have quieted Pauling. Instead it had just the opposite effect. In a new book, *Vitamin C and the Common Cold and the Flu* (Pauling, 1976), he claimed that the medical tests had used totally inadequate amounts of Vitamin C. Massive doses, a gram or more per day, were needed. And he went further. He asserted that Vitamin C in such large doses helps with the treatment of hepatitis, mumps, measles, polio, viral pneumonia, viral orchitis, herpes, and influenza. He proposed mechanisms by which Vitamin C does its job, both as a substance that mops up free chemical radicals in cells and as a component of a cancer-cell inhibiting chemical called PHI. He also pointed out that there was no danger of a vitamin overdose, since excess Vitamin C is harmlessly excreted from the body.

Again, the medical control experiments were done. Again, Pauling's claims were denied, and dismissed. That is where the question stands today. Books have been written, proposing Vitamin C as a practical panacea for all ailments. Others have totally rejected all its beneficial effects. The use of large doses of Vitamin C remains a scientific heresy.

However, in discussing this subject with scientists, I find that a remarkably high percentage of them take regular large doses of Vitamin C. Perhaps it is no more than a vote of solidarity for a fellow-scientist. Perhaps it is a gesture of respect toward Linus Pauling, who died in August 1994 in his ninety-fourth year.

Or perhaps it is more the attitude of the famous physicist Niels Bohr. He had a horseshoe nailed up over the doorway of his country cottage at Tisvilde, for good luck. A visitor asked if Bohr, a rational person and a scientist, really believed in such nonsense. "No," said Bohr, "but they say it works even if you don't believe in it."

13.5 Minds and machines. In Chapter 10, we described the extraordinary advance of computers. The first ones, in the 1940s, were used for straightforward calculations, of tables and payrolls and scientific functions. Since then the applications have spread far beyond those original uses. Computers today perform complex algebra, play chess and checkers better than any human, control power generating plants, keep track of everything from taxes to library loans to airplane reservations, check our spelling and the accuracy of our typing, and even accept vocal inputs that may soon make typing unnecessary.

Given a suitable program, no human effort of calculation and record-keeping seems to be beyond computer duplication. This raises natural questions: Is every function of the human mind really some form of computer program? And at some time in the future, will computers be able to "think" as well as humans?

To most of the scientists represented in Chapter 10, the answer to these questions is an unequivocal "Yes." Our thought processes operate with just the same sort of logic as computers. Our brains are, as Marvin Minsky said, "computers made of meat." The field of Artificial Intelligence, usually abbreviated as AI, seeks to extend the range of those functions, once thought to be uniquely powers of the human mind, that computing machines are able to perform. The ultimate goal is a thinking and "self-conscious" computer, aware of its own existence exactly as we are aware of ours.

That ultimate goal seems far off, but not unattainable—unless a distinguished mathematician, Roger Penrose, is right. In 1989, he offered a radically different proposal. This is the same Penrose that we met in Chapter 2. He is the Rouse-Ball Professor of Mathematics at Oxford University, and a man with a reputation for profound originality. Over the past thirty years he has made major contributions to general relativity theory, to numerical analysis, to the global geometry of space-time, and to the problem of tiling the plane with simple shapes. His work is highly diverse, and it is characterized by ingenuity and great geometrical insight. More important, many of his results are surprising, finding solutions to problems that no one else had suspected might exist, and stimulating the production of much work by other investigators. Even his harshest critics admit that Roger Penrose is one of the world's great problem solvers. He cannot be dismissed outright as a crank, or as an intellectual lightweight.

What then, does he propose?

In a book that was a surprising best-seller, *The Emperor's New Mind* (Penrose, 1989), he claimed that some functions of the human brain will never be duplicated by computers that develop along today's lines. The brain, he asserts, is "non-algorithmic," which means that it performs some functions for which no computer program can be written.

This idea seems like perfect scientific heresy, and it was received with skepticism and even outrage by many workers in the field of AI and computer science (for a brief summary, see *How the Mind Works* [Pinker, 1997]). For one thing, prior to this book, Penrose was very much one of their own kind. Now he seemed like a traitor.

Marvin Minsky even called Penrose a "coward," which is a perplexing term since it takes a lot of nerve to propose something so far out of the scientific mainstream.

What does Penrose say that is so upsetting to so many? In *The Emperor's New Mind*, he argues that human thought employs physics and procedures quite outside the purview of today's AI and machine operations. The necessary physics is drawn from the world of quantum theory. In Penrose's words, "Might a quantum world be *required* so that thinking, perceiving creatures, such as ourselves, can be constructed from its substance?" (Penrose, 1989).

His answer to that question is, yes, such a quantum world is required. To see the direction of his argument, it is necessary to revisit what was said in Chapter 2 about quantum theory.

In the quantum world, a particle does not necessarily have a well-defined spin, speed, or position. Rather, it has a number of different possible positions or speeds or spins, and until we make an observation of it, all we can know are the probabilities associated with each possible spin, speed, and position. Only when an observation is made does the particle occupy a well-defined state, in which the measured variable is precisely known. This change, from undefined to well-defined status, is called the "collapse of the quantum mechanical wave function." This is a well-known, if not well-understood, element of standard quantum theory.

What Penrose suggests is that the human brain is a kind of quantum device. In particular, the same processes that collapse the quantum mechanical wave function in subatomic particles are at work in the brain. When humans are considering many different possibilities, Penrose argues that we are operating in a highly parallel, quantum mechanical mode. Our thinking resolves and "collapses to a thought" at some point when the wave function collapses, and at that time the many millions or billions of possibilities become a single definite idea.

This is certainly a peculiar notion. However, when quantum theory was introduced in the 1920s, most of its ideas seemed no less strange. Now they are accepted by almost all physicists. Who is to say that in another half-century, Penrose will not be equally accepted when he asserts, "there is an essential *non*-algorithmic ingredient to (conscious) thought processes" and "I believe that (conscious) minds are *not* algorithmic entities"?

Meanwhile, almost everyone in the AI community (who, it might be argued, are hardly disinterested parties) listens to what Penrose has to say, then dismisses it as just plain wrong. Part of the problem is Penrose's suggestion as to the mechanism employed within the brain, which seems bizarre indeed.

As he points out in a second book, *Shadows of the Mind* (Penrose, 1994), he is not the first to suggest that quantum effects are important to human thought. Herbert Fröhlich, in 1968, noted that there was a high-frequency microwave activity in the brain, produced, he said, by a biological quantum resonance. In 1992, John Eccles proposed a brain structure called the *presynaptic vesicular grid*, which is a kind of crystalline lattice in the brain's pyramidal cells, as a suitable site for quantum activity.

Penrose himself favors a different location and mechanism. He suggests, though not dogmatically, that the quantum world is evoked in elements of a cell known as microtubules. A microtubule is a tiny tube, with an outer diameter of about 25 nanometers and an inner diameter of 14 nanometers. The tube is made up peanut-shaped objects called *tubulin dimers*. Each dimer has about ten thousand atoms in it. Penrose proposes that each dimer is a basic computational unit, operating using quantum effects. If he is right, the computing power of the brain is grossly underestimated if neurons are considered as the basic computing element. There are about ten million dimers per neuron, and because of their tiny size each one ought to operate about a million times as fast as a neuron can fire. Only with such a mechanism, Penrose argues, can the rather complex behavior of a single-celled animal such as a paramecium (which totally lacks a nervous system) be explained.

Penrose's critics point out that microtubules are also found elsewhere in the body, in everything from livers to lungs. Does this mean that your spleen, big toe, and kidneys are to be credited with intelligence?

My own feeling is that Penrose's ideas sounded a lot better before he suggested a mechanism. The microtubule idea feels weak and unpersuasive. Like the Wizard of Oz, the theory was more impressive when it was hidden away behind the curtain.

My views, however, are not the issue. Is Penrose wrong, destined to be remembered as a scientific heretic? Or is he right, and a true prophet?

It is too soon to say. But if he proves to be right, his ideas will produce a huge change in our conceptions of physics and its relation to consciousness. More than that, the long-term future of computer design will become incredibly difficult.

With the latter point in mind, we might paraphrase Bertrand Russell. He said of Wittgenstein's theories, as we can say of Penrose's: "Whether they are true or not, I do not know; I devoutly hope that they are not, as they make mathematics and logic almost incredibly difficult."

Meanwhile, I am waiting for a story to appear making use of Penrose's extraordinary claim that we are controlled by quantum processes within the brain's microtubules.

13.6 Diseases from space. In the late 1970s, two respected scientists proposed an interesting and radical new theory (Hoyle and Wickramasinghe, 1977): Certain diseases are often not carried from one person to another by the usually accepted methods, sometimes mutating as they go to become other strains of the same infection; instead, the diseases arrive on Earth from space, and the observed variations arise there.

In the words of Fred Hoyle and Chandra Wickramasinghe (1977), the joint proposers of the theory:

"In *Diseases from Space* we shall be presenting arguments and facts which support the idea that the viruses and bacteria responsible for the infectious diseases of plants and animals arrive at the Earth from space."

They support their contention on biochemical grounds,

and also from statistical evidence on the spread of influenza in Britain.

The same two workers also suggest that life itself did not develop on Earth. It was borne here, as viruses and bacteria.

Again in their words: "Furthermore, we shall argue that apart from their harmful effect, these same viruses and bacteria have been responsible in the past for the origin and evolution of life on the Earth. In our view, all aspects of the basic biochemistry of life come from outside the Earth."

Where, then, did life originally develop? Hoyle and Wickramasinghe give their answer: It arose naturally in that great spherical collection of comets known as the Oort Cloud, which orbits far beyond the observable solar system. They argue that conditions for the spontaneous generation of life were far more favorable there than they were on Earth, three and a half billion years ago when life first appeared here.

This idea is not totally original with them. Early this century, the Swedish chemist Svante August Arrhenius proposed that life is widespread in the universe, being constantly diffused from one world to the next in the form of spores. The spores travel freely through space, now and again reaching and seeding some new habitable world (Arrhenius, 1907). Hoyle and Wickramasinghe, while not accepting this *panspermia* concept totally, and substituting viruses and bacteria for Arrhenius's spores, do claim that life was brought to Earth in a similar fashion.

Hoyle and Wickramasinghe also deny that many epidemics of infectious disease are spread by person-to-person contact or through intermediate carriers (such as lice and mosquitoes). They claim that influenza, bubonic plague, the common cold, and smallpox all originate in the fall of clouds of infecting spores (bacteria or viruses) from space, and are mainly spread by incidence from the air.

This sounds, on the face of it, somewhat unlikely. No one has ever observed a virus or a bacterium present in space, or arriving from space. However, the reaction of the medical community went far beyond polite skepticism. The new idea was ignored or vilified as preposterous, and it was treated as a true scientific heresy.

Why was the reaction of the medical establishment so strong?

First, there was a question of qualifications. Not as scientists, where the credentials of both proposers are impeccable. Hoyle is one of the world's great astrophysicists, a man who has made profound contributions to the field, and Wickramasinghe is a well-known professor. However, neither Hoyle nor Wickramasinghe is a physician or a microbiologist. They were astronomers, operating far outside their own territory.

Second, the presently accepted idea for disease transmission was itself once a scientific heresy. It took three hundred years for the notion that tiny organisms can invade the human body and cause infections to change from wild surmise to scientific dogma. Such a theory, so hard-won, is not readily abandoned. Thus, in 1546, Girolamo Fracastoro proposed a germ theory of disease. In his book *De Contagione*, he suggested three modes of transmission: by direct contact, indirectly through such things as clothing, and through the air. He was generally ignored, if not actively ridiculed.

The situation changed only in the late eighteenth century, when scientists were able to verify the existence of bacteria by direct observation with the microscope. And it was not for almost a hundred years more, until the second half of the nineteenth century, that Louis Pasteur and Robert Koch put the matter beyond question when they isolated the specific bacterial agents that cause anthrax, rabies, cholera, and tuberculosis, and used inoculation to protect against several of them.

The modern picture of disease transmission then appeared to be complete, and it is not far from Fracastoro's original ideas. Contagious diseases spread from person to person. Some call for personal contact, like syphilis. Some can be transmitted through the air, like the common cold. Some diseases, like malaria, require the action of an intermediate organism such as a mosquito; and some, like trichinosis, can be transmitted by the ingestion of infected food. However, all communicable diseases have one thing in common: they originate somewhere on the surface of the Earth, and they are carried by terrestrial organisms.

This leads at once to the third and perhaps the biggest objection to Hoyle and Wickramasinghe's theory: there is overwhelming direct evidence for the conventional means of disease transmission. Even if the new theory were to prove right in part, it cannot be the whole story. Thus, the rapid spread of bubonic plague through Europe in the fourteenth century, and the almost instantaneous and devastating effects of smallpox on native American Indians when it was brought by Europeans in the early sixteenth century, owe nothing at all to space-borne spores. The attacks were too sudden and the timing too coincidental. These diseases ran riot in populations which had no previous exposure to them, and therefore lacked protective antibodies against them.

Ultimately, then, the main argument against the theory offered by Hoyle and Wickramasinghe may not be that it is ridiculous, or biologically unfounded, or in some way impossible. It is that it is not *necessary*, since the established notions of disease propagation seem quite sufficient to explain everything that we see, and are required for that explanation.

Until today's theories prove inadequate, or there is better evidence for the new theory, the idea that diseases arrive from space will remain what it is today: a scientific heresy.

13.7 Cold fusion. On March 23, 1989, a press conference was held at the University of Utah. The organizers of the conference stated that they had managed to initiate and sustain a nuclear fusion reaction. That announcement astonished the world, for several unrelated reasons.

First, the use of a press conference is not the normal method for announcement of a scientific discovery. Scientists have a well-defined procedure for doing this: the discovery is described in enough detail for others to know what has been done, and to begin the process of verification; in the case of an important discovery,

where precedent may be important, a brief note is sent
to the appropriate scientific journal and preprints are sent
to professional colleagues. Today, the preprint often takes
the form of an e-mail letter. Scientists do not choose a
press conference as the appropriate mechanism to reveal
their discoveries. Those discoveries do not, as this one
did, take over newspaper headlines around the world and
lead to wild speculation in certain metals.

The second reason for astonishment was the nature
of the claimed discovery itself. Nuclear fusion is well-
known to science. The fusion of hydrogen to helium is
the main process that allows the sun and stars to shine.
Here on Earth, nuclear fusion makes possible the
hydrogen bomb. Large experimental facilities in this
country and elsewhere have spent billions of dollars over
the past forty years, trying to tame the violent fusion
of the hydrogen bomb to permit a controlled release of
energy. Nuclear fusion looks like the Holy Grail of endless
and clean energy production, but the experimental
equipment needed is large and complex, and employs
temperatures of tens or hundreds of millions of degrees—
hotter than the center of the sun.

By contrast, the nuclear fusion described in the Utah
press conference takes place at room temperature—"cold"
fusion—and calls for only the simplest of means. All that
is needed is a beaker of "heavy" water and a palladium
electrode. Heavy water is water in which the normal
hydrogen atoms have been replaced by deuterium, a rare
but well-known heavier form of hydrogen (see Chapter
5). Heavy water is naturally present in ordinary water,
at a concentration of about one part in six thousand.
Palladium is a steely-white metal, also rather rare but
well-known and widely available.

The final surprise in the Utah announcement was the
identity of the two scientists given credit for the discovery.
Martin Fleischmann had a distinguished career in England
before retiring as an emeritus professor from the University of Southampton and beginning the work in Utah.
He is a Fellow of Britain's most prestigious scientific
group, the Royal Society, and has been described by
colleagues as "more innovative than any other

electrochemist in the world." Stanley Pons had been a student under Fleischmann at Southampton, before becoming the prolifically productive head of the University of Utah chemistry department. Both men thus had excellent credentials—as chemists. Nuclear fusion, however, is a problem calling for knowledge not of chemistry but of *physics*. It requires an understanding of the processes by which the nuclei of atoms can be combined.

Physicists as a group often do not have the highest regard for chemistry, which they consider as messy and unsystematic and more like cooking than science. It was, therefore, unusually satisfying to chemists and galling to physicists when Fleischmann and Pons, using the simplest of means, seemed to have made the whole expensive business of conventional nuclear fusion experiment, as performed by physicists, seem irrelevant.

Fleischmann and Pons had an explanation for the way their results had been achieved. At first sight that explanation seemed very plausible. It has been known for generations that palladium has a high natural affinity for hydrogen. A palladium rod, placed in a hydrogen atmosphere, will absorb up to nine hundred times its own volume of hydrogen. It will do the same thing if heavy hydrogen is used in place of ordinary hydrogen. According to Fleischmann and Pons, the palladium electrode would absorb heavy hydrogen from the heavy water, and within the palladium the heavy hydrogen nuclei would be so close to each other that some of them would fuse. The result would be helium and heat. Neutrons, an elementary particle present in the heavy hydrogen, would be released as a by-product. Fleischmann and Pons reported seeing significant heat, more than could possibly be produced by chemical processes, and a small number of neutrons. All of this happened at room temperature, in a beaker no bigger than a peanut butter jar.

The press conference did not give details of the process, so other groups had trouble at first either confirming or denying the claimed results. It took several months before a coherent picture emerged. When the dust settled, the verdict was not in favor of Fleischmann and Pons. Some other groups observed a few, a very few, neutrons, barely more than the normal background level. Others reported excess heat, but no neutrons, and again it was nowhere near what had been claimed by the Utah group.

Why didn't Fleischmann and Pons seek confirmation from those other groups, before they made their announcement? To some extent, they did, and they were still in the process of doing so. However, great pressure to make that announcement prematurely, and to do it through a press conference, came not from the two chemists but from officials at the University of Utah. The university administrators could see an enormous profit potential if the cold fusion claims held up. That potential would only be realized if patents were granted and the Utah claim to precedence recognized. It must have seemed like a good bet, at least to the officials: the reputation of two professional chemists, against possible multiple billions of dollars of gain for the university.

Today, the bet appears to be over. Fleischmann and Pons were the losers. They still insist that their original results are correct, and continue their research not in Utah but in France, with private funding. However, few other reputable scientists believe they will find anything valuable.

Even at the very beginning, there were basic physical reasons to discount the "cold fusion" claim. The number of neutrons observed was far too small, by a factor of billions, to be consistent with the claimed heat production. Real fusion would produce huge numbers of neutrons, enough to be fatal to anyone in the same room as the beaker with its palladium electrode.

Many people continue to believe ardently in cold fusion. I do not, though some new phenomenon—not fusion—may be there. And I must say, I feel a great deal of sympathy for Pons and Fleischmann. They were pushed by university administrators into making the premature announcement of results.

Had they followed a more conventional route, the results might have been very different. The obvious parallel is in the area of high temperature superconductivity. In 1986, Müller and Bednorz produced the first ceramic superconductors. Such things were "impossible" according to conventional theories. But when experiment and theory disagree, theory must change. Müller and Bednorz won the 1987 Nobel Prize for physics.

Were Pons and Fleischmann robbed of similar fame by the actions of others? Possibly. However, martyrdom is not enough to make a theory correct. Today, cold fusion remains as scientific heresy.

13.8 No Big Bang. The standard model of cosmology sees the Universe as beginning in a primordial, highly condensed fireball that has been expanding ever since. Such a model explains the recession of the galaxies, the 2.7 Kelvin microwave background radiation, and the relative abundance of the elements, particularly hydrogen and helium. Each of these independent phenomena seems to provide powerful observational evidence.

Critics of the Big Bang point out that the theory does not explain the mystery of the missing matter, nor how galaxies formed in the first billion years of an originally smooth universe. The nature of quasars is also open to question.

It is one thing to object to a theory. It is another to offer a viable alternative. What do we have that might replace the Big Bang cosmology?

There are two independent groups critical of the Big Bang theory. The first is led by Fred Hoyle, Halton Arp, and Geoffrey Burbidge. Arp has done considerable observational work on quasars, showing that some of them with large red shifts seem to be physically connected to galaxies displaying much smaller red shifts. If this is the case, then the whole redshift-distance correlation falls apart. And Hoyle, back in the late 1940s, along with Thomas Gold and Hermann Bondi, proposed an alternative to the Big Bang known as the "continuous creation" or "steady state theory." (Hoyle, more recently, says that the word "steady" was a bad choice. He, Gold, and Bondi meant only to indicate that the rate of recession of the galaxies does not change with time, as is the case with Big Bang cosmology.) The original form of the steady-state theory, however, had other problems; observations did not support the independence of galactic age with distance that it predicted.

Hoyle, along with the Indian astronomer Narlikar, has developed a new and different version of the steady-state theory, this one consistent with ideas of cosmological inflation needed also by the Big Bang theory. Instead of a single, one-time Big Bang, however, Hoyle and Narlikar posit a large number of small acts of creation, arising from vacuum fluctuations and suffering rapid expansions or "inflations."

Hoyle also proposes another mechanism to explain the microwave background radiation, although in a sense, he hardly needs to. Scientists long ago, knowing nothing about an expanding universe or the recession of the galaxies, had calculated the temperature of open space. Charles Guillaume, in a paper published in 1896, calculated a temperature of 5.75 Kelvin. Eddington in 1926 estimated the temperature as 3 Kelvin. Early proponents of the Big Bang, by contrast, believed the temperature ought to be higher, anywhere from 7 to 30 Kelvin.

Hoyle and his associates argue that the microwave background radiation stems from a well-known process called *thermalization*. All that is happening, they say, is that the light from stars is being scattered to longer and longer wavelengths by its interactions with metallic "whiskers" (length-to-width ratios of 1:100,000) seeded throughout interstellar and intergalactic space by supernova explosions. In Hoyle's words, Nature is an "inveterate thermalizer," and the process will continue until actual stellar radio sources dominate—which is in the microwave region.

The second group of Big Bang critics began with Hannes Alfven, a Swedish Nobel Prize winner in Physics. His work has been continued by Anthony Peratt and Eric Lerner, and the resulting theory is usually termed "plasma cosmology."

Plasma cosmology has its own proposed mechanism

for explaining the 2.7 Kelvin background radiation. It is based on a theory proposed in 1989 by Emil Wolf. The "Wolf shift" shows how light passing through a cloud of gas is shifted in frequency toward the red end of the spectrum. This effect and the thermalization effect (both of which may be operating) throw question on the recession rates of the galaxies.

The plasma cosmology group also argues that most of the matter in the universe is not electrically neutral, but charged—free electrons, or positively charged nuclei. Since this is the case, electromagnetism, rather than gravity, is the controlling force. Alfven, making this assumption, concluded that sheets of electric current must crisscross the Universe. Interacting with these, plasma clouds would develop a complex structure and complex motions. Alfven predicted that the universe would display a cellular and filamentary nature over very large scales.

At the time, the universe seemed to be smooth at such scales, and his ideas were not accepted. Evidence of "walls" and "voids" and galactic super-clusters did not appear until the mid-1980s. Today, the supporters of the Big Bang are hard pressed to explain what Alfven's theory establishes in a natural way.

Finally, there is the question of the abundance of the elements. Both the Hoyle school and the plasma cosmology school have pointed out that, according to the Big Bang's own equations, the abundances of four light nuclei—hydrogen, deuterium, helium, and lithium—must all be linked. If the helium abundance of the universe, today, is below 23 percent (as observation indicates) then there will be more deuterium than observed. In fact, there will be eight times as much. On the other hand, if the density of the universe is high enough to avoid producing too much deuterium in the very early days, then there is not enough helium now. It should be more than 24 percent. And finally, when we put lithium and deuterium together into the picture, the necessary helium abundance comes out as over 25 percent.

In other words, juggle the Big Bang theory as you like, you cannot come up with a version that provides the observed amounts of the three substances.

We see that the three mainstays of Big Bang theory are subject to alternative explanations or open to question. So where does that leave us?

Well, today the Big Bang remains as standard dogma; anything else, steady-state or continuous creation or plasma cosmology, is still scientific heresy.

However, I can't help feeling that Big Bang theory has some major problems. Its proponents, if they are at all sensitive, must feel the winds of change blowing on the back of their necks. In science, that is usually a healthy sign.

13.9 Free energy. Energy from nothing, free electricity drawn from the air. Who could resist it? I include this more for fun than anything else, and because I experienced the matter at first hand.

It began in February 1996, with a telephone call from Arthur Clarke in Sri Lanka: "There's going to be a demonstration on March 5th, at Union Rail Station in Washington, D.C. A group claims to have a way of generating free electricity. I can't go. Can you?"

I could, and did. The East Hall of Union Station had been rented for the occasion, with an overflow room upstairs that offered a real-time video of the live performance. The event was scheduled to begin at 7:30. I went along early, and made a point of talking with the group (the Columbus Club) who rented out the space. One man talked of the event as a "show"; i.e., an entertainment. A lady responsible for registration had been told to expect about nine hundred people. My estimate of the actual turnout is maybe one hundred, at the beginning. By the time I left, roughly at ten o'clock, no more than fifty were left.

The man who did most of the talking was tall, American, dark-haired, and very experienced in presentation. He spoke for two and a half hours, without notes. He had a disarming manner, and constantly referred to himself as more lucky than clever. In fact, because he disdained technical knowledge, it was hard to question him about technical matters.

He began by describing a heat pump that seemed

quite conventional and comprehensible. The "low-temperature phase change" technology is exactly that employed in a refrigerator, with freon as the material that is cycled. He described his heat pump as better than anyone else's, and that may be true. It would require a good deal of study to prove otherwise. Less plausible was the method by which the pump was designed. He said he came across it by accident, taking an evaporator, a compressor, and a condenser, of rather arbitrary sizes, dumping freon into them, and finding that the result performed better than anything else available—by a factor of three. Because he said he was not technical, answers to some important questions were not forthcoming. Did he vary the parameters of his system? Or, how does he know that his design is the best that there is?—a claim that he made.

However, the heat pump claim was completely testable. If he had stopped there (we were maybe forty minutes into his two and a half hour presentation) I would have been favorably impressed.

I had more trouble with the next, nontechnical statements that he made. He asserted that he had built this device in the early 1980s and made 50 million dollars in 18 months. The electric companies then drove him out of business, made three attempts on his life, threatened his associates, and in some way not altogether clear arranged for him to serve two years in jail. Time not wasted, he says, because he had 17 new energy production ideas while locked up.

Ignoring all that, I was disturbed by three technical aspects of the next part of the presentation. First, there was a continuing confusion between energy *storage* and energy *production*. Second, he moved smoothly from the statement that he had a very efficient heat pump, which I could accept, to a statement that he had a pump that produced a net energy output—in other words, the efficiency was more than a hundred percent. Third, every method of producing energy that I know of relies on the existence of a thermal gradient, usually in the form of a hot and a cold reservoir. These two seemed to be muddled in his discussion, with the heat flow going in the wrong direction. In other words, he took energy from his cold reservoir and put it into his hot reservoir. That's a neat trick.

We move on. Next he introduced a new device, which he termed the *Fischer engine*. This, as he describes it, is a steam (or, if you prefer, freon) engine which uses super-heated water (or freon) under high pressure and requires no condenser. His explanation of how it works left me unsatisfied. However, I believe that the Fischer engine could be a genuine advance. He stated that "the Carnot cycle is not a major concern." Since the Carnot cycle represents an ideal situation in which all processes are reversible, his statement is equivalent to saying that the second law of thermodynamics is not a major concern, at least to him. It is to me.

After this, things became less comprehensible. He coupled his heat pump with the Fischer engine, and asserted that the result would be more powerful than an internal combustion engine of the same size, would never need gasoline (or any other fuel), would not need any oil changes, and would run for 400,000 miles before it wore out.

If some of those statements seem remarkable, during the final hour several much more striking ones were offered. He declared that he knew five different ways to produce free electricity. He stated that he had seen a working anti-gravity machine. Finally, he again insisted that his desire to offer the world (or at least the United States) free, unlimited, pollution-less energy was being thwarted by the utility companies.

In summary, it was a fascinating evening. However, it seems a little premature to sell that Exxon or BP stock. It also proves to me that entrepreneurs are still out there, trolling the deep waters for sucker fish.

13.10 Wild powers. So far in this chapter we have discussed what might be called "offshore science." The ideas might be heretical, and an occasional proponent might suggest lunacy or charlatanism, but they live within the general scientific framework. Now we are heading for deep water.

Let us begin with a quandary. What can you say about something for which every scientific test has turned up no evidence, but which 90 percent of the people—maybe it's 99 percent—believe?

I am referring to the "wild powers" of the human mind. A short list of them would have to include telepathy, clairvoyance, prescience, psychokinesis, divination, dowsing, teleportation, reincarnation, levitation, channeling, faith healing, hexing, and psychics.

In addition to these, another group of widely-held beliefs involves aliens: abduction by aliens, sightings of alien spacecraft, rides in alien spacecraft, impregnation by aliens, and—a central element of the movie *Independence Day*—aliens who landed on Earth, only to have their existence concealed by the U.S. Government.

I know at least one person who believes in each of these things, often while rejecting many of the others as ridiculous. People who pooh-pooh the idea of, say, UFOs, will accept that humans, in times of stress, can communicate over long distances with close family members—and I am not referring to telephone calls. At least one United States president, Ronald Reagan, permitted astrology to play a part in his administration. Another's wife, Hillary Clinton, may have tried to channel Eleanor Roosevelt, though she later claimed that it was just a game. I know several trained scientists who believe that the government is covering up knowledge of alien landings on Earth—although they acknowledge that the government has been singularly inefficient in hiding other secrets.

This is a book about science. Rather than engage in pro and con arguments for the hidden powers of the human mind, or the presence or absence of aliens, I will say only this: good science fiction stories have been written using every item on my list. Some of them are among the best tales in the field. Thus, Alfred Bester used telepathy in *The Demolished Man* (Bester, 1953), and teleportation in *The Stars My Destination* (Bester, 1956). Robert Heinlein had aliens taking over humans in *The Puppet Masters* (Heinlein, 1951). Theodore Sturgeon employed a variety of wild powers in *More Than Human* (Sturgeon, 1953), as did Frank Herbert in *Dune* (Herbert, 1965). Zenna Henderson, in her stories of The People, used aliens and wild powers to great effect (Henderson, 1961, 1966).

These writers took unlikely ingredients, and used them to produce absolute classics. Feel free to go and do thou likewise.

13.11 Beyond the edge of the world. Finally, let's take a trip right to the edge of the world and off it, with the *Kidjel Ratio*. I feel fairly confident that this has never been used in a science fiction story, and it's never going to be used in one of mine, so it's all yours.

It's not often that a revolutionary new scientific advance makes its first appearance in the *Congressional Record*. But here we go:

> From the Congressional Record
> of the U.S. House of Representatives,
> 3 June, 1960:
> The Kidjel Ratio—A New Age in
> Applied Mathematics and Arts
> Extension of Remarks of
> Hon. Daniel K. Inouye of Hawaii.

"Mr. Speaker, Hawaii's wealth in human resources has once again proved to be unlimited. The ingenuity and pioneering spirit of its citizens have given to the world a new and practical system of solving a multitude of problems in the important fields of applied mathematics, art, and design.

"The Kidjel ratio is now being used to great advantage in more than 40 related activities in the world of architecture, engineering, mathematics, fine arts and industrial arts . . .

"Academically speaking, the Kidjel ratio also led to the discovery of the solutions of the three famous 2,500 year-old so-called impossible problems in Greek geometry, popularly known as:

"First. Trisecting the angle—dividing an angle into three equal parts.

"Second. Squaring the circle—constructing a square equal in area to a given circle.

"Third. Doubling the cube—constructing a cube, double in volume to that of a given cube with the use of compasses and unmarked ruler only . . ."

What is the Kidjel ratio? I have no idea. But as to the three mathematical advances cited, all three have been proved mathematically impossible with the restriction imposed by the Greeks who originally proposed them (solution must be done by a geometrical construction of a finite number of steps, using compasses and unmarked ruler only). The first and third were shown to be impossible by about 1640, when Descartes realized that their solutions implied the solution of a cubic equation, which cannot be done using ruler and compasses only. The second was disposed of in 1882, when Ferdinand Lindemann proved that π is a "transcendental" number, not capable of being expressed as the solution of any algebraic equation.

What is impossible to mathematicians is apparently simple enough for the U.S. House of Representatives. Moreover, the Kidjel ratio amendment is not without precedent. Here we have another fine example:

"A bill for introducing a new mathematical truth, and offered as a contribution to education to be used only by the State of Indiana free of cost by paying any royalties whatever on the same . . ."—House Bill No. 246, introduced in the Indiana House on January 18, 1897.

Section 1 of the bill continues:

"Be it enacted by the General Assembly of the State of Indiana: It has been found that a circular area is to the square on a line equal to the quadrant of the circumference, as the area of an equilateral rectangle is to the square on one side. The diameter employed as the linear unit according to the present rule in computing the circle's area is entirely wrong . . ."

In other words, the value of π, the ratio of the circumference of a circle to its diameter, and one of the most fundamental numbers in mathematics, was not what mathematicians believed it to be. The value of π is an infinite decimal, 3.141592653 . . . which can be approximated as closely as desired, but not given exactly. However, a correct and exact value was promised to the Indiana legislature, thanks to the efforts of an Indiana physician, Dr. Edwin J. Goodwin.

This piece of nonsense would probably have become law, except for the timely arrival at the State Capitol of Professor C.A. Waldo, a member of the mathematics department of Purdue University, there on quite other business. He was astonished to find the House debating a piece of mathematics, with a representative from eastern Indiana saying, "The case is perfectly simple. If we pass this bill which establishes a new and correct value of π, the author offers without cost the use of this discovery and its free publication in our school textbooks, while everyone else must pay him a royalty."

Professor Waldo managed to educate the senators. They voted to postpone the bill indefinitely on its second reading. The State of Indiana learned a lesson, and passed up a wonderful opportunity to become the laughing-stock of the mathematical world. But as Senator Inouye proves, no lesson lasts forever.

13.12 One last heresy. Not all science fiction stories have to be serious. They can, if you prefer it, be ridiculous. Here, as an example, is one which plays games with a few basic ideas of physics. I leave it to the reader to pick up the references to theories and people. I will only add that all the formulas quoted in the story are correct.

THE NEW PHYSICS: THE SPEED OF LIGHTNESS, CURVED SPACE, AND OTHER HERESIES

Listwolme is a small world with a thin but permanently cloudy atmosphere. The inhabitants have never seen the stars, nor become aware of anything beyond their own planet. There is one main center of civilization which confined itself to a small region of the surface until about a hundred years ago, when an industrial revolution took place. For the first time, rapid transportation over substantial areas of the planet became possible.

Orbital velocity at the surface of Listwolme is less than two kilometers a second. The meetings of the Listwolme Scientific Academy following the development of high-velocity surface vehicles are chronicled below. The highlights of those meetings were undoubtedly the famous exchanges between Professor Nessitor and Professor Spottipon.

The first debate: In which Professor Nessitor reveals the curious results of his experiments with high-speed vehicles, and proposes a daring hypothesis.

Nessitor: As Members of the Academy will recall, a few months ago I began to install sensitive measuring devices aboard the *Tristee Two*, the first vehicle to move at a speed more than ten times that of a running *schmitzpoof*. The work was not easy, because it was first necessary to suppress all vibration induced by the car's contact with the surface.

One month ago we achieved the right combination of smooth suspension and vibration damping. It was with some excitement that I placed one of our instruments, a sensitive spring balance, within the vehicle and we began steadily to increase our speed. As you may have heard, there have been reports of "feeling light" from the drivers of these cars when they go at maximum velocity.

Fellow scientists, those feelings are no illusion! Our instruments showed a definite decrease in load on the balance as our speed was increased. There is a relationship between *weight* and *motion*!

(As Nessitor paused, there was a murmur of surprise and incredulity around the great hall. Professor Spottipon rose to his feet.)

Spottipon: Professor Nessitor, your reputation is beyond question. What would arouse skepticism from another in your case is treated with great respect. But your statement is so amazing that we would like to hear more of these experiments. For example, I have heard of this "lightening" effect at high speeds, but seen no quantitative results.

Were your balances sensitive enough to measure some relation between the lightness and the speed?

Nessitor (triumphantly): With great precision. We measured the weight shown on the balance at a wide variety of speeds, and from this I have been able to deduce a precise formula between the measured weight, the original weight when the vehicle was at rest, and the speed of movement. It is as follows.

Here Professor Nessitor went to the central display screen and sketched on it the controversial formula. It is believed that this was the first time it had ever appeared to public view. In the form that Nessitor used, it reads:

$$\text{(Weight at speed v)} = \text{(Rest weight)} \times (1 - v^2/c^2)$$

When the formula was exhibited there was a silence, while the others examined its implications.

Spottipon (thoughtfully): I think I can follow the significance of most of this. But what is the constant, c, that appears in your equation?

Nessitor: It is a velocity, a new constant of nature. Since it measures the degree to which an object is lightened when it moves with velocity v, I suggest that the basic constant, c, should be termed the "speed of lightness."

Spottipon (incredulously): You assert that this holds anywhere on Listwolme? That your formula does not depend on the *position* where the experiment is conducted?

Nessitor: That is indeed my contention. In a series of experiments at many places on the surface, the same result was obtained everywhere, with the same velocity, "c." It is almost four times as fast as our fastest car.

(There was a long pause, during which Professor Spottipon was seen to be scribbling rapidly on a scribe pad. When he had finished his face bore a look of profound inspiration.)

Spottipon: Professor Nessitor, the formula you have written has some strange implications. You assert that there is a lightening of weight with speed across the surface. This we might accept, but you have not taken your formula to its logical limit. Do you realize that there must be a speed when the weight *vanishes*? When v=c, you have a situation where an object does not push at all on the balance! Worse than that, if v *exceeds* your "speed of lightness" you would calculate a *negative* weight. If that were true, a car moving at such a speed would fly completely off the surface. You would have created the long-discussed and arguably impossible "flying machine."

Nessitor (calmly): As Professor Spottipon has observed with his usual profound insight, the speed of lightness is a most fundamental constant. My interpretation is as follows: since it is clearly ridiculous that an object should have negative weight, the formula is trying to tell us something very deep. It is pointing out that *there is no way that an object can ever exceed the speed of lightness*. The speed that we can deduce from these experiments, c, represents the ultimate limit of speed that can ever be attained.

(Sensation. The assembled scientists began to talk among themselves, some frankly disbelieving, others pulling forth their scribe pads and writing their own calculations. At last a loud voice was heard above the general hubbub.)

Voice: Professor Nessitor! Do you have any name for this new theory of yours?

Nessitor (shouting to be heard): I do. Since the effects depend only on the motion *relative* to the ground, I suggest the new results should be termed the *principle of relativity*. I think that . . .

(Professor Nessitor's next comments were unfortunately lost in the general noise of the excited assembly.)

Six months passed before Professor Nessitor appeared again at a meeting of the Academy. In those months, there had been much speculation and heated argument, with calls for more experiments. It was to an expectant but still skeptical audience that the Professor made his second address.

Nessitor: Distinguished colleagues, last time that I was here there were calls for proof, for some fundamental basis for the formula I presented to you then. It was to answer those calls that I embarked, four months ago, on a new set of experiments with the *Tristee Two* vehicle. We had installed a new instrument on board our car. It measures distances very accurately, and permits the car's course to be controlled to an absolutely straight line. For it had occurred to me to ask the question, if velocity and weight are so closely linked, could it be that distance itself depends on some unknown factors?

Spottipon (somewhat irritably): With all due respect, Nessitor, I have no idea what you mean by such a statement. Distance is distance, no matter how fast you traverse it. What could you hope to find? I hoped that you would have repeated the experiments on speed and weight.

Nessitor: My esteemed colleague, please have patience. Permit me to tell you what happened. We set the *Tristee Two* to travel a long distance at various speeds. And indeed, we confirmed the speed-weight relation. At the same time, we were measuring the distance traveled. But in performing this experiment we were moving longer linear distances over the surface of Listwolme than any other scientific group had ever done.

I therefore decided to conduct an experiment. We traveled a long distance in a certain direction, accurately measuring this with our new instrument. Then we made a half turn and proceeded far along this new line, again measuring distance all the way. Finally, we headed straight back to our original starting point, following the hypotenuse of the triangle and measuring this distance also.

Now, we are all familiar with the Sharog-Paty Theorem that relates the lengths of the sides of a right-angled triangle.

(Nessitor went to the central display panel and scribed the famous Sharog-Paty relation: $c^2 = a^2 + b^2$. There was a mutter of comments from behind him.)

Impatient voice from the audience: Why are you wasting

our time with such trivia? This relation is known to every unfledged child!

Nessitor: Exactly. But it is not what we found from our measurements! On long trips—and we made many such—*the Sharog-Paty relation does not hold*. The further we went in our movements, the worse the fit between theory and observation.

After some experiment, I was able to find a formula that expresses the true relation between the distances a, b, and c. It is as follows.

(Nessitor stepped again to the display panel and wrote the second of his famous relations, in the form:

$$\cos(c/R) = \cos(a/R) \times \cos(b/R)$$

There was more intense study and excited scribbling in the audience. Professor Spottipon alone did not seem to share in the general stir. His thin face had gone pale, and he seemed to be in the grip of some strong private emotion. At last he rose again to his feet.)

Spottipon: Professor, old friend and distinguished colleague. What is "R" in your equation?

Nessitor: It is a new fundamental constant, a distance that I calculate to be about three million paces.

Spottipon (haltingly): I have trouble saying these words, but they must be said. In some of my own work I have looked at the geometry of other surfaces than the plane. Professor Nessitor, the formula you have written there already occurs in the literature. It is the formula that governs the distance relations *for the surface of a sphere*. A sphere of radius R.

Nessitor: I know. I have made a deduction from this—

Spottipon: I beg you, do not say it!

Nessitor: I must, although I know its danger. I understand the teachings of our church, that we live on the Great Plain of the World, in God's glorious flatness. At the same time I cannot ignore the evidence of my experiments.

(The Great Hall had fallen completely silent. One of the recording scribes dropped a scribe pin in his excitement and received quick glares of censure. It was a few seconds before Nessitor felt able to continue. He stood there with head bowed.)

Nessitor: Colleagues, I must say to you what Professor Spottipon with his great insight realized at once. The distance formula is identical with that for distances on a sphere. My experiments suggest that *space is curved*. We live not on a plane, but on the surface of an immense sphere.

(The tension crackled around the hall. The penalty for heresy—smothering in live toads—was known to all. At last Professor Spottipon moved to Nessitor's side and placed one hand on his shoulder.)

Spottipon: My old friend, you have been overworking. On behalf of all of us, I beg you to take a rest. This "curved space" fancy of yours is absurd—we would slide down the sides and fall off!

(The hall rang with relieved laughter.)

Spottipon: Even if our minds could grasp the concept of a curved space, the teachings of the Church must predominate. Go home, now, and rest until your mind is clearer.

(Professor Nessitor was helped from the stage by kind hands. He looked dazed).

For almost a year, the Academy met without Nessitor's presence. There were rumors of new theories, of work conducted at white heat in total seclusion. When news came that he would again attend a meeting, the community buzzed with speculation. Rumors of his heresy had spread. When he again stood before the assembly, representatives of the Church were in the audience. Professor Spottipon cast an anxious look at the Churchmen as he made Nessitor's introduction.

Spottipon: Let me say how pleased we are, Professor Nessitor, to welcome you again to this company. I must add my personal pleasure that you have abandoned the novel but misguided ideas that you presented to us on earlier occasions. Welcome to the Academy!

Nessitor (rising to prolonged applause, he looked nervous but determined): Thank you. I am glad to be again before this group, an assembly that has been central to my whole working life. As Professor Spottipon says, I have offered you some new ideas over the past couple of years, ideas without fundamental supporting theory. I am now in a position to offer a new and far more basic approach. *Space is curved, and we live on the surface of a sphere!* I can now prove it.

Spottipon (motioning to other scientists on the stage): Quick, help me to get him out of here before it's too late.

Nessitor (speaking quickly): The curvature of space is real, and the speed of lightness is real. But the two theories are not independent! The fundamental constants c and R are related to a third one. You know that falling bodies move with a rate of change of speed, g, the "gravitational constant." I can now prove that there is an exact relation between these things, that $c^2 = g \times R$. To prove this, consider the motion of a particle around the perimeter of a circle . . .

(The audience was groaning in dismay. Before Nessitor could speak further, friends were removing him gently but firmly from the stage. But the representatives of the church were already moving forward.)

At his trial, two months later, Professor Nessitor recanted all his heretical views, admitting that the new theories of space and time were deluded and nonsensical. His provisional sentence of toad-smothering was commuted to a revocation of all leaping privileges. He has settled quietly to work at his home, where he is writing a book that will be published only after his death.

And there were those present at his trial who will tell you that as Nessitor stepped down from the trial box he whispered to himself—so softly that the words may have been imagined rather than heard—"But it *is* round."

CHAPTER 14
The End of Science

Maybe it is because we are at the turn of the century, facing the new millennium. Maybe it is because there has been no obvious big breakthrough for a couple of decades. Maybe global pessimism is the current fad. For whatever reason, several recent books have suggested that the "end of science" may be in sight.

Their titles betray the direction of their thinking: *Dreams of a Final Theory: The Scientist's Search for the Ultimate Laws of Nature* (Weinberg, 1992); *The End of Physics: The Myth of a Unified Theory* (Lindley, 1993); *The End of Science: Facing the Limits of Knowledge and the Twilight of the Scientific Age* (Horgan, 1996).

These all suggest, in the case of the last book with considerable relish, that the great moments of science have all occurred; that scientists are now on a road of diminishing returns; and that the next hundred years will offer nothing remotely comparable to the discoveries of the last few centuries.

Steven Weinberg is a physicist, and a great one. He would very much like to see a "theory of everything" in his lifetime. That does not mean that everything will then have been explained, only that the most basic underpinnings of everything, which he sees as the laws of fundamental physics, will have been established. He recognizes that there will be more questions to be answered, and perhaps many discoveries in other branches of science that will come to be regarded as absolutely radical and basic. But physics is nearing its final form.

David Lindley and John Horgan are both editors, at *Science* magazine and *Scientific American* respectively. Lindley, after a careful review of the development of physics since the end of the last century, disagrees with Weinberg. He concludes that the "theory of everything will be, in precise terms, a myth. A myth is a story that makes sense within in its own terms . . . but can neither be proved nor disproved." Scientists in pursuit of a final theory are then like dogs chasing an automobile. What will they do with it if they catch it?

Horgan takes a broader approach. He interviewed scores of eminent scientists who have made major contributions in their diverse fields. In the end, his conclusion is that, whether or not we are approaching a final theory, we are at any rate at the end of all discoveries of the most fundamental kind. The scientists of future generations will mainly be engaged in mopping-up operations.

This tune may sound familiar. It has been heard before, notably at the end of the nineteenth century. Here is Max Planck, recalling in 1924 the advice given to him by his teacher, Philipp von Jolly, in 1874: "He portrayed to me physics as a highly developed, almost fully matured science . . . Possibly in one or another nook there would perhaps be a dust particle or a small bubble to be examined and classified, but the system as a whole stood there fairly secured, and theoretical physics approached visibly that degree of perfection which, for example, geometry has had already for centuries."

Is it more plausible now than it was then, that the end of science is in sight? And if so, what does it mean for the future of science fiction?

As Sherlock Holmes remarked, it is a capital mistake to theorize before one has data. Let us examine the evidence.

First, let us note that because Horgan's scientists are already recognized major figures, most of them are over sixty. None is under forty, many are over seventy, a few are well into their eighties, and several have died in the two years since the book was published. Although everyone interviewed seems as sharp as ever, there is an element of human nature at work which Horgan himself recognizes and in fact points out. Gregory Chaitin, in a discussion with Richard Feynman, said he thought that science was just beginning. Feynman, a legend for open-mindedness on all subjects, said that we already know the physics of practically everything, and anything that's left over is not going to be relevant.

Chaitin later learned that at the time Feynman was dying of cancer. He said, "At the end of his life, when the poor guy knows he doesn't have long to live, then I can understand why he has this view. If a guy is dying he doesn't want to miss out on all the fun. He doesn't want to feel that there's some wonderful theory, some wonderful knowledge of the physical world, that he has no idea of, and he's never going to see it."

We are all dying, and anyone over seventy is likely to be more aware of that than someone twenty-five years old. But the latter is the age, particularly in science, where truly groundbreaking ideas enter the mind. If we accept the validity of Chaitin's comments, Horgan's interviews were foreordained to produce the result they did. Science, as perceived by elderly scientists, will always be close to an end.

As Arthur Clarke has pointed out, when elderly and distinguished scientists say that something can be done, they are almost always right; when they say that something cannot be done, they are almost always wrong. Fundamental breakthroughs, carrying us far from the scientific mainland, are, before they take place, of necessity unthought if not unthinkable. That is the philosophical argument in favor of the idea that we are not close to the end of progress. There is also a more empirical argument. Let us make a list of dates that correspond to major scientific events. Lists like this tend to be personal; the reader may choose to substitute or add milestones to the ones given here, or correct the dates to those of discovery rather than publication.

1543: Copernicus proposes the heliocentric theory displacing Earth from its position as the center of the universe.

1673: Leeuwenhoek, with his microscopes, reveals a whole new world of "little animals."

1687: Isaac Newton publishes *Principia Mathematica*, showing how Earth and heavens are subject to universal, calculable laws.

1781: Herschel discovers Uranus, ending the "old" idea of a complete and perfect solar system.

1831: Michael Faraday begins his groundbreaking experiments on electricity and magnetism.

1859: Darwin publishes *The Origin of Species*, dethroning Man from a unique and central position in creation.

1865: Mendel reports the experiments that establish the science of genetics.

1873: Maxwell publishes *A Treatise on Electricity and Magnetism*, giving the governing equations of electro-magnetism.

1895–7: Röntgen, Becquerel, and J.J. Thomson reveal the existence of a world of subatomic particles.

1905: Einstein publishes the theory of special relativity.

1925: The modern quantum theory is developed, primarily by Heisenberg and Schrödinger.

1928: Hubble discovers the expansion of the universe.

1942: The first self-sustaining chain reaction is initiated by Fermi and fellow-workers in Chicago.

1946: The first digital binary computer is built by Eckert and Mauchly.

1953: Crick and Watson publish the structure of the DNA molecule.

1996: Evidence is discovered of early life-forms on Mars.

Is there a pattern here? The most striking thing about this list of dates and events might seem to be the long gap following 1953, since the discovery of Martian life is still highly tentative. We have not seen so long a hiatus for more than a hundred years.

But is the gap real? In the 1830s, Faraday's experiments on electricity were considered fascinating, but hardly something likely to change the world. In 1865, scarcely anyone knew of Mendel's experiments—they lay neglected in the *Proceedings of the Brünn Society for the Study of*

Natural Science for twenty years. And in 1905, only a small handful of people realized that the relativity theory offered a radically new world-view. (Max Born, later one of Einstein's closest friends, wrote: "Reiche and Loria told me about Einstein's paper, and suggested that I should study it. This I did, and was immediately deeply impressed. We were all aware that a genius of the first order had emerged." Born, however, was himself a genius. It takes one to know one.)

It also takes a long time to accept ideas that change our basic perception of reality. Remember that Einstein was awarded the Nobel Prize in 1921 mainly for his work on the photoelectric effect, and not for the theory of relativity. That was still considered by many to be controversial.

Will posterity record the year that you read this book as an *annus mirabilis*, the marvelous year when the defining theory for the next centuries was created?

Am I an optimist, if I find that suggestion easier to believe than that we, in this generation, are seeing for the first time in scientific history the wall at the edge of the world? Humans are often guilty of what I call "temporal chauvinism." It takes many forms: "We are the first and last generation with both the resources and the will to go into space. If we do not do it now, the chance will be lost forever." "We are the final generation in which the Earth is able to support, in comfort, its population." "We are the last generation who can afford to squander fossil fuels." "After us, the deluge."

I believe that science, science new and basic and energetic, has a long and distinguished future, for as far as human eye can see. And I believe that science fiction, which as science draws on contemporary developments but which as literature draws on all of history, will play an important role in that future.

Certainly, we can envision and write about times as bleak and grim as you could choose; but we can also imagine better days, when our children's children may regard the world of the late twentieth century with horror and compassion, just as we look back on the fourteenth century in Europe.

Science fiction fulfills many functions; to entertain, certainly—otherwise it will not be read—but also to instruct, to stimulate, to warn, and to guide.

That is science fiction at its best, the kind that you and I want to read and write. I see no reason why any of us should settle for less.

APPENDIX:
Science Bites

I offered the warning back on Page 1, in the very first sentence: "You are reading an out-of-date book." Science marches on, exploring new territories and expanding older ones every week.

I knew this, but I didn't think I could do anything about it. Fortunately, I was wrong. Just about the time that *Borderlands of Science* was reaching the book stores, I was invited to begin a weekly science column for distribution to newspapers and other media (especially on-line outlets).

There was only one catch. The columns would have to be very short, "science bites" rather than science articles; six or seven hundred words, rather than the six or seven *thousand* that I am used to. I squirmed at the prospect—what could I possibly say in six hundred words?—but I couldn't deny the logic of the argument. The world speeds up, attention spans are down, so science bites won't catch the fish; all they can do is set the bait, so that an interested reader can follow up with longer articles or books. The good news was that I could write on any subject I liked. And the title of the newspaper column? What else but *The Borderlands of Science*.

Here, then, is a little bait, a couple of dozen of those brief articles. All were written after the main body of *Borderlands of Science* was complete. The Borderland has moved a little farther out. And if you want to see how it is still moving, go online to www.paradigm-tsa.com for more of the weekly columns.

A.1. The ship jumped over the moon. No matter what *Star Trek* and *Babylon 5* may tell you, moving objects around in space is a tricky business. The Space Shuttle can sometimes do it, provided that it doesn't have to go after anything more than about 300 miles up. The in-space fix of the Hubble Space Telescope was a spectacular success. But if a satellite gets into trouble in a high orbit, thousands of miles from Earth, it's usually beyond saving.

That's the way it looked in December 1997 when the failure of a rocket booster sent a Hughes communications satellite into the wrong orbit. The spacecraft was supposed to sit at a fixed longitude, 22,300 miles above the Pacific Ocean. Instead it traced a looping, eccentric path, varying widely in its distance from the surface of the Earth.

Time to give up? It seemed tht way. From its changing position, the satellite could not deliver communications and television in Asia. And although the spacecraft had a small rocket of its own on board, there was not enough propellant to move it directly into the correct orbit. Insurers examined the situation and declared the satellite a total loss. In April 1998, they gave ownership back to Hughes, saying in effect, "Here's a piece of junk way out in space. It's all yours, do what you like with it."

Hughes engineers did, through a surprising and spectacular idea: Although the satellite's rocket was not big enough to force it directly into the right orbit, it could float the spacecraft out to the Moon. Once there, the lunar gravity field might be used to change the orbit of the satellite. In effect, the spacecraft would get a "free boost" from the Moon, stealing a tiny amount of Luna's vast orbital energy to modify the satellite's own speed and direction.

The first swing around the Moon was made in May 1998, after which the spacecraft came looping back in toward Earth. The orbit still wasn't right, so another small rocket firing and a second lunar swing-by was made three weeks later. This time the satellite returned close to its desired orbit. A final firing of the rocket engine in mid-June, 1998, did the trick. The spacecraft now sat in a 24-hour circular orbit, just as originally planned, going around the Earth at the same speed as the world turns on its axis.

It sits there now, drifting a few degrees north and south of the equator every day while remaining close to a constant longitude. Known as HGS-1, it is working perfectly and ready for use in global communications. More than that, HGS-1 serves as a tribute to human ingenuity. When a space mission in trouble had officially been declared dead, engineers down here on Earth "repaired" it without ever leaving their chairs. Perhaps even more impressive, to anyone who remembers the first disastrous attempts to launch an American satellite: this round-the-moon space shot didn't rate television coverage or a newspaper headline. We've come a long way in forty years.

A.2. Future cars. I'm a writer, so there's a chance my works will live on. But I agree with Woody Allen, I don't want to live forever through my works. I want to live forever by not dying.

That presents certain problems. If I—and you—don't die, we will certainly get older. Sixty years from now, without some spectacular medical advance, none of us will look or feel young. The retina of a 75-year-old has only 10 percent of the sensitivity of an 18-year-old. By age 75, we are at least a little deaf (particularly the men). The range of mobility of our neck and shoulders is down, and our reaction times are slower. However, if today's 70- and 80-year-olds are anything to go by, we'll insist on one thing: we want to drive our own cars. It's part of our independence, our ability to look after ourselves.

Let's put that together with another fact, and see where it takes us: People are living longer, and the US population is getting older. In 1810, there were only about 100,000 people over 65. By 1880 it was close to 2 million. By 1960, 16 million, and today it's over 30 million. In

2030 it will be near 60 million. And most of these aging people—remember, that's you and me—will still want to drive their cars.

We will need help, and fortunately we will get it. Auto manufacturers who study ergonomics—the way that people operate in particular situations—are already taking the first steps.

The driver doesn't see or hear too well? Fine. The car provides a "virtual reality" setting. Actual light levels outside the car will be changed, so that what the driver sees compensates for loss of visual sensitivity. The driver will receive an enlarged field of view without having to turn very far, so as to compensate for decreased head and neck mobility. The speed of reaction of the driver will be improved using servo-mechanisms, just as today the strength of a driver is augmented by power steering.

These are all, relatively speaking, easy. They can be done today. Most older drivers already wear glasses. We simply replace them with goggles that present virtual reality views of the surroundings. The driver should hardly notice, except to remark how much clearer everything seems.

At the same time, the car will do more things for itself: monitoring engine temperatures, stresses, loads, and driving conditions. Rather than presenting this information in the "old-fashioned" way, through dials and gauges, the car's computer will report only when something is outside the normal range.

More complex, and farther out in time, comes the involvement of the car's control systems in real-time decision making. Here, the automobile not only senses variables from the environment, it also *interprets* the inputs, draws conclusions, and recommends actions (ACCIDENT FOUR MILES AHEAD; SUGGEST YOU LEAVE FREEWAY AND TAKE ALTERNATE ROUTE. SHALL I MAKE ADJUSTMENT AND ESTIMATE NEW ARRIVAL TIME?). Or, in emergency, the car's computer will initiate action without discussion. A human cannot react in less than a tenth of a second. A computer can react in a millisecond. The difference, at 60 miles an hour, is about 10 feet—enough to matter.

These changes to the automobile are more than probabilities; I regard them as future certainties. My job, and yours, is to be around long enough to enjoy them.

A.3. Making Mars. A hundred years ago, Mars was in the news. H.G. Wells had just published his novel, *The War of the Worlds*, and everyone seemed convinced that there must be life on the planet. Astronomers even thought they had seen through their telescopes great irrigation "canals," showing how water was moved from pole to pole.

Today, Mars is a hot topic again. Some scientists believe they have found evidence of ancient Mars life in meteorites flung from there to Earth. Others say, forget the ancient past. Mars is the place for life in the *future*—human life. Let's go there, explore, set up colonies, and one day transform Mars so that it is right for people. Mars has as much land area as Earth; it could be a second home for humanity.

Sounds great. NASA ranks Mars high on the list of its priorities. Can we make another planet where humans can live, work, raise families, and have fun? How easy is it to change Mars so it is more like our own planet?

In a phrase, it's mighty tough. Mars has plenty of land. What it does not have are three things we all take for granted: air, water, and heat. Making Mars more like Earth—"terraforming" the planet—requires that we provide all three.

Heat should be the easy one. We can load the thin atmosphere of Mars with CFC's, "greenhouse gases" currently in disfavor on Earth because they contribute to global warming. As the temperature rises, solid carbon dioxide held in the Mars polar ice caps will be released into the air, trapping more sunlight and adding to the warming process. The Mars atmosphere, currently only about one percent as dense as ours, will thicken. At the same time, the temperature will rise enough for water, held below the surface as permafrost, to turn to liquid as it is brought to the surface. Recent estimates suggest enough water on Mars to provide an ocean three hundred feet deep over the whole surface.

When the warming process is complete Mars will have heat, water, and air. Unfortunately, that air will be mostly carbon dioxide. Humans and animals can't breathe that—but growing plants rely on it, taking it in and giving out oxygen. The key to making breathable air on Mars is through the import of Earth plant life, genetically engineered to match Mars conditions.

Now for the catch. If we started today, how long would it take to transform Mars to a place where humans could survive on the surface? In the best of circumstances, assuming we use the best technology available today and make this a high-priority project, the job will take four or five thousand years.

That's as much time as has elapsed since the building of the Egyptian pyramids. The technology available to our far descendants is likely to be as alien and incomprehensible to us as computers and genetic engineering and space travel would have been to the ancient Egyptians. Maybe we ought to wait a while longer before we start changing Mars.

A.4. Close cousins: How near are we to the great apes?

A visit to the monkey house at the zoo is a sobering experience. We stare in through the bars. Looking right back at us with wise, knowing eyes is someone roughly our shape and size, standing like us on two feet, perhaps pointing at us with fingers much like our own and apparently laughing at us. He bears an uncanny resemblance to old Uncle Fred. Maybe we should look twice to make sure who is on the right side of the bars.

It is easy to believe that of all the creatures in the animal kingdom, the chimps, gorillas, and orangutans are nearer to humans than any other. The question is, how close?

A generation ago, we could offer only limited answers. We were different species, because inter-breeding was impossible. As for other similarities and differences, they had to be based on the comparison of muscle and bone structure and general anatomy.

Now we have new tools for the comparison of species. The complete genetic code that defines a gorilla is

contained in its DNA, a gigantic long molecule organized
into a number of long strands called chromosomes.
Moreover, every cell of a gorilla (or a human) contains
the DNA needed to describe the complete animal. Given
a single cell from a chimp and a cell from a human,
we can take the DNA strands and do a point-by-point
comparison: the structure is the same here, different there.
The extent to which the two DNA samples are the same
is a good measure of the closeness of the two species.

This analysis has been performed, and the results are
breathtaking. Humans and chimps share more than
ninety-eight percent of their DNA. Each of us is, in an
explicit and meaningful way, less than two percent away
from being a chimp.

The same exercise, carried out with DNA from orangutans, shows that humans are rather less closely related
to them. As we consider other animals, everything from
a cheetah to a duck to a wasp, we find that our intuitive ideas are confirmed. The differences between our DNA
and those of other creatures steadily increases, as the
species become more obviously "different" from us in form
and function. DNA analysis tells us that we are more like
every other mammal than we are like any bird, and we
are more like every bird than we are like any insect.

We can use this and other information to estimate how
long ago different species diverged from each other.
Humans and chimps have a common ancestor which
lived roughly five million years ago. Humans and gorillas
diverged at much the same time, as did chimps and
gorillas from each other. We and the orangutans parted
ancestral company farther in the past, about twelve to
fifteen million years ago.

Five million years may sound like a long time, but there
has been life on Earth for more than three and a half
billion years. We and the great apes separated very
recently on the biological time scale, and we really are
close cousins. It should be no surprise that we feel an
odd sense of family recognition when we meet them.

A.5. Breathing space. How many can Earth hold? Stuck
in rush hour traffic on a hot day, you sometimes wonder:

Where did all these people come from? You may also mutter to yourself, Hey, it wasn't like this when I was a kid.

If you do say that, you'll be right. It wasn't. There are more people in the world today than ever before. Next year there will be more yet.

Here are some sobering numbers: At the time of the birth of Christ, the world population was around two hundred million. By 1800 that number had climbed to one billion. Two billion was reached about 1930; three billion in 1960; four billion in 1975; five billion in 1988. By 2000 the population topped six billion.

Not only more and more, but faster and faster. People are living longer and the old equalizers, famine and plague, seem largely under control. War remains, but the two great conflicts of the twentieth century did little to slow the growth in population. It took a hundred and thirty years to add the second billion, only twelve years to add the sixth. With any simple-minded projection for the next century apparently zooming to infinity, we have to ask, how long can this go on? Also, where will all the new people *fit*? Not, we hope, on the commuter routes that we use.

A glance at a population map at first seems reassuring. Most of the world still looks empty. A second glance, and you realize why. Of the Earth's total land area, one fifth is too cold to grow crops, another fifth is too dry, a fifth is too high, and another seventh has infertile soils. Only about a quarter of the land is good for farming. Empty areas of the planet are empty for good reason.

That can change, and is already changing. Sunlight is available everywhere, regardless of how high or cold the land. Plants are remarkably efficient factories for converting sunlight, water, and carbon dioxide into food. By genetic engineering, we are producing new crop varieties with shorter growing seasons and more tolerance of cold, drought and high salinity. In the future, it seems certain that areas of the globe now empty will be able to fill with high-yield food crops, and then with people.

According to projections, they will have to. Population estimates for the year 2050 range from eight to twelve billion, for 2100 from ten to fifteen billion. Even these numbers are nowhere near the limit. Provided that we can produce the food (and distribute it), the Earth can easily support as many as twenty billion people—almost four times as many as we have today.

The question you may ask, sitting in your car on a freeway that has become one giant parking lot, is a different one. Sure, if we struggle and squeeze, maybe we can handle four times the present population. But do we want to? How many people is *enough*?

A.6. The inside view. Last week I had to go for a CAT scan of my abdomen. It was nothing special, more a loss of dignity than anything else. But it was still an indignity, still uncomfortable, still *invasive* (I had to swallow a barium milk shake that tasted like glue). Most people hate the standard medical routines. We don't like lying semi-naked on a slab, or having blood taken, or the prospect of the insertion of mechanical devices into veins, sinuses, bronchial tubes, urethras, and other personal apertures.

How might we improve all this? Before we get to that, let's recall the past—and things far worse than anything that will happen to me next week.

Until a hundred years ago a doctor had few tools to examine a patient's interior. A stethoscope to listen to body processes, a thermometer to measure body temperature, a spatula to examine tongue and throat, and that was about it. The rest of the diagnosis relied on palpation (feeling you), tapping, external symptoms such as ulcers and rashes, and the appearance of various waste products. If all those failed, the dreaded next step could be "exploratory surgery," with or without anesthetics.

And then, as though by magic, in the last decade of the nineteenth century a device came along that could actually *see* inside a human body. True, the X-ray was better at viewing bones and hard tissue than organs and soft tissue, but it was an enormous step forward.

The X-ray was the first "modern" tool of diagnosis.

Since then we have developed a variety of other methods for taking an inside look: the CAT scan, the MRI, and ultrasonic imaging. They permit three-dimensional images of both hard and soft tissues. Used in combination with the injection of special materials, they allow the operation of particular organs or the flow of substances through the body to be studied as they happen. The radioactive tracers used for this purpose are one undeniably positive fall-out of the nuclear age.

Hand in hand with the images goes chemical diagnosis. Today, blood samples can provide a doctor with information about everything from liver function to diabetes to urinary infection to rheumatoid arthritis to AIDS to the presence of the particular bacterium (Helicobacter pylori) responsible for most stomach ulcers. Any tiny skin sample is sufficient to permit a DNA analysis, which can in turn warn of the presence of certain hereditary diseases and tendencies.

We have come a long way in a hundred years. What about the next hundred?

Completely non-invasive diagnosis, with superior imaging tools and with chemical tests that can operate without drawing blood, will become available in the next generation. The chemical tests will use saliva and urine samples, or work through the skin without puncturing it. For the digestive system, we may swallow a pill-sized object, which will quietly and unobtrusively observe and report on the whole alimentary canal as it passes along it.

No more upper GI exams, no more sigmoidoscopies. Our grandchildren will regard today's invasions (drawing blood, taking bone marrow samples, or inserting objects into the body) the way we think of operations without anesthetics: part of the bad old days of the barbarous past.

Or even the uncomfortable indignity of last week.

A.7. Attack of the killer topatoes. Genetically modified vegetables are in the news. Should you eat these "genemods" (GMs, in England) or should you avoid them at all costs?

Let's back up a bit. Your wife may not like it when you say that her brother is a louse. You explain to her, that's not really an insult. At the most basic level, down where it matters, a louse is not that different from a human being.

How so? Your brother-in-law and the louse are made up of cells. So are you. In the middle of almost every cell is a smaller piece called the cell nucleus. Inside that, even smaller, are long strings of material called chromosomes. It's your chromosomes that decide what you are like, while the louse's decide what it is like.

When you examine a human or a louse chromosome in detail they look remarkably similar. And it's not just the louse and your brother-in-law. It's everything you can think of, from snails to snakes to Susan Sarandon. All our chromosomes are made of the same basic stuff; and that stuff is what makes each of us, physically, what we are.

This suggests a neat idea. Suppose we have a variety of tomato that is very tasty and productive, but suffers from tomato wilt. We also have a wilt-resistant type of potato. The immunity is carried in a particular part of a potato chromosome (let's call it a "gene"). If we could snip just that gene out, and insert it into the tomato chromosome, we might be able to create a new something—call it a "topato"—that produces great tomatoes and doesn't suffer from wilt.

We are not quite that smart yet, although with some foods we are well on the way. Do you believe you have never eaten genetically modified foods? Then take a look at the boxes and containers in your kitchen. See if they contain soy (most of them do). Today nearly half of the US soybean crop is a genetically engineered variety. Crops with genetic modification can have a higher yield, or an extra vitamin, or greater tolerance for weedkillers, bacteria, or salty soil. The variations are endless, and the topato I'm describing is a real possibility.

Should we worry about all this? I think we should be careful. We are making things that never existed in Nature. Maybe a "Frankenstein tomato" could have new properties that never occurred to us when we were making it.

On the other hand, plant and animal breeders have been playing this game, through cross-breeding, for thousands of years. There was no such thing as a nectarine or a loganberry before a human developed them, but we eat them quite happily. Even if we hadn't made such things, nature has a way of trying so many different combinations that they might occur naturally in time.

My only concern is that we may be, as usual, in a bit too much of a hurry. We are duplicating in a few years a process that in nature would normally take millions; and we, unlike nature, have to explain our mistakes.

A.8. Twinkle, twinkle? "It's not the things we don't know that causes the trouble, it's the things we know that ain't so." I love that quote and I wish I'd said it first, but Artemus Ward beat me to it by about a hundred and fifty years.

Less than fifty years ago, one thing every astronomer "knew" was that there was a limit to what a telescope could see when looking out into space. If you made a telescope's main lens or mirror bigger and bigger, it would collect more and more light but the degree of detail of what you saw would not increase. The limiting mirror or lens size is quite small, about ten inches, and beyond that you will get a brighter but not a sharper image.

The problem is nothing to do with the telescope's design or manufacture. The spoiler is the Earth's atmosphere, which is in constant small-scale turbulence. The moving air distorts the path of the light rays traveling through it, so that instead of appearing as a steady, sharp image, the target seems to be in small, random motion. The nursery rhyme has it right. When the target looks small, like a star, it will twinkle; when it is larger and more diffuse, like a planet or galaxy, fine detail will be blurred.

Twenty years ago that was the end of the story. If you wanted highly detailed images of objects in space, you had to place your telescope outside the Earth's atmosphere. That idea led to the orbiting Hubble Space

Telescope, whose wonderful images have appeared on every TV channel and in every magazine. The Hubble pictures are far more detailed than any obtained by a telescope down here on Earth, even though the size of the Hubble's mirror, at 94 inches, is much smaller than the 200-inch mirror at Mount Palomar. The only road to detailed images of astronomical objects was surely the high road, through telescopes placed in orbit.

This "fact" turned out to be one of the things we know that ain't so. About fifteen years ago, a small group of scientists working on a quite different problem for the Strategic Defense Initiative ("Star Wars" to most people) came up with the idea of aiming a laser beam upward and measuring the way that its path was distorted in the atmosphere. Knowing what happened to the laser beam, the focus of the observing telescope mirror could be continuously (and rapidly) changed, so as to compensate for the changes in light path. The procedure, known as "adaptive optics," was tried. It worked, spectacularly well. Today, ground-based telescopes are obtaining images of a crispness and clarity that a generation ago would have been considered impossible.

What else do we "know" that can't be done with ground-based telescopes today? Well, the Earth's atmosphere completely absorbs light of certain wavelengths. If we want to learn what is happening in space at those wavelengths, we still need orbiting telescopes. I certainly believe that is true. On the other hand, it may be just one more thing I know that ain't so.

A.9. Are you a cyborg?

At the turn of the millennium, I get asked one question over and over: What's going to happen to *us*? How will we change, as humans, when science and technology advance over the years and the centuries?

The only honest answer is, I don't know; but I am willing to stick my neck out and make a prediction in one specific area: we will all become, more and more, cyborgs.

A cyborg is a human being, changed to improve or restore body functions by the addition or replacement of man-made parts. Almost everyone reading this is already a cyborg in one or more ways. Are you wearing eyeglasses or contact lenses? Do you have dental fillings, or a crown on a tooth? Are you perhaps wearing a hearing aid, or a pacemaker, or is one of your knee, hip, or shoulder joints artificial? Has part of a vein or artery been replaced by a plastic tube?

If your answer to any of these questions is yes, then you are part cyborg. Admittedly, these are cyborg additions at the most primitive level, but we already have the technology to make much more versatile and radical changes to ourselves.

Let's consider a few of the easy ones. First, we can make an artificial eye lens containing miniature motors, sensors, and a tiny computer. The lens will adapt, just like a human eye lens, to changes in light levels and in the distance of the object being viewed. Nearsightedness, far-sightedness and astigmatism will become history. As the human retina ages, or light levels become low, the lens can also boost the contrast of scenes to compensate. Everyone will have eyes like a hawk, able to see with great clarity, and eyes like a cat, able to see well in near-dark. Last night, driving an unfamiliar winding road through heavy snow, I would have given a lot for a pair of these future eye lenses.

At first, of course, such things will cost a lot; millions of dollars for the prototypes. But, like hand calculators or cameras, once they are in mass production prices will fall dramatically. The main cost will be the one-time installation charge.

Suppose that your eyes are excellent, and you have no need for cyborg eyes. What about your hearing? Today's hearing aids, despite the claims made for them, are rotten. They don't give directional hearing, and they can't separate what you want to hear from background noise. The next generation of hearing aids will also contain tiny computers. They will be invisibly small, provide full stereo directional hearing, and boost selected sound frequencies as necessary. They too will be expensive at first, but manufacturing costs will drop until they are cheap enough to throw away rather than repair.

Your ears and eyes are in fine working shape, you say, so you don't need cyborg help? Very well. Here are a few other third millennium optional additions. You choose any items that appeal to you.

Peristalsis control, to provide perfectly regular bowel habits. A sleep regulator, which can be set to make you fall asleep or awake according to your own preferred schedule. A general metabolic rate regulator, boosting or lowering body activity levels to match the situation (or the level of a partner; we probably all know couples who wage constant war over setting the thermostat). A blood flow controller, solving any possible problems of male impotence. A vocal cord monitor, which adjusts your rough shot at a note so you sing exactly in tune. Built-in computer chips, to provide instant answers to arithmetic and logical questions of all kinds.

If this list worries you, and you say, isn't there a danger that devices like this will sometimes be abused or misused? I reply, can you think of any piece of technology that sometimes isn't?

A.10. "You've got a virus." Sometimes I think that viruses were created mainly to benefit the medical profession.

You're not feeling well, and you go to see your doctor. After an examination and a test or two, she says, "You're sick all right. You have an infection. But it isn't a bacterial infection, it's a viral infection. So there's no point in giving you antibiotics. Just go home and take it easy until you feel better." Meaning, "We're not quite sure what's wrong with you, but we do know we can't give you anything to cure it."

Are viruses and bacteria really so different? On the face of it, they have a lot in common. They exist in large numbers everywhere, some forms serve as the agents for disease, and they are too small to be seen without a microscope. On closer inspection, however, viruses are much more mysterious objects than bacteria.

First, although both are tiny, viruses are orders of magnitude smaller. The largest known bacterium is relatively huge, a bloated object as big as the period at the end of this sentence. Bacteria are complete living organisms, which reproduce themselves given only a supply of nutrients.

By contrast, a virus is a tiny object, often less than a hundred-thousandth of an inch long. It is no more than a tiny piece of DNA or RNA, wrapped in a protein coat, and it cannot reproduce at all unless it can find and enter another organism with its own reproducing mechanism. It is different enough from all other life forms that some biologists argue that viruses are not really alive; certainly, they do not fit into any of the known biological kingdoms.

The way in which a virus reproduces is highly ingenious. First, it must find and penetrate the wall of a normal healthy cell, often with the aid of a little tail of protein that serves as a kind of corkscrew or hypodermic syringe. Once inside, the virus takes over the cell's own reproducing equipment. It uses that equipment to make hundreds of thousands of copies of itself, until the chemical supplies within the cell are used up. Then the cell wall bursts open to release the viruses, which go on to repeat the process in another cell. Viruses are, and must be, parasitic on other life forms. They are the ultimate Man Who Came to Dinner, who does not leave until he has eaten everything in the house, and also killed his host.

This explanation of what a virus is and does leads to a bigger mystery: Since a virus totally depends for its reproduction on the availability of other living organisms, how did viruses ever arise in the first place?

Today's biology has no complete answer to this question. However, it seems to me that the only plausible explanation is that viruses were once complete organisms, probably bacteria with their own reproducing mechanisms. They found it advantageous to invade other cells, perhaps to rob them of nutrients. As time went on, the virus found that it could get by with less and less of its own cellular factories, and could more and more use the facilities of its host. Little by little the virus dispensed with its cell wall and its nutrient-producing facilities, and finally retained only the barest necessities needed to copy itself. What we see today is the result of a long process of evolution, which could perhaps more appropriately be called devolution. The end result is one of nature's most perfect creations, reproduction reduced to its absolute minimum.

The virus is a lean, mean copying machine. It may be a comfort to remember this, the next time that you are laid low by what your doctor describes as a viral infection. And we can take greater comfort from the fact that, as our understanding increases, twenty years from now we should have "viral antibiotics" to tackle viruses and have us back on our feet within 24 hours.

A.11. Accelerating universe. For the past year the astronomers of the world have been in a state of high excitement. Observations of supernovas—exploding stars—billions of light-years away suggest a surprising result: the universe, which since the 1920s has been known to be expanding, is not simply expanding; it's *accelerating*. Distant galaxies are not only receding from us, they are flying away faster and faster.

Since these events are taking place at distances so great as to be almost unimaginable, the natural reaction to the new observations might well be, so what? How can things so remote have any possible relevance to human affairs here on Earth?

To answer that question, we need to explain why it is so surprising for far-off galaxies to be moving away increasingly fast. The place to start is with the "standard model" of the universe, the mental picture of the cosmos that scientists have been developing and testing for the past seventy years. According to that model, our universe began somewhere between twelve and twenty billion years ago, in a "Big Bang" that sent all parts of that original tightly-compressed universe rushing away from each other. We have to point out that it is not that other parts of the universe are receding *from us*, which would imply we are in some special position. *All* parts are running away from each other. And the first evidence of this expansion was provided in 1929 by Edwin Hubble, after whom the Hubble Space

Telescope is named. All observations since then confirm his result.

Will the expansion continue forever, or it will it stop at some future time? That question proved difficult to answer. The force of gravity operates on every galaxy, no matter how far away, and it acts to pull them all closer together. Given enough material in the universe, the expansion might one day slow down and even reverse, with everything falling back together to end in a "Big Crunch." Or, with less density of material, the expansion might go on forever, with the force of gravity gradually slowing the expansion rate. But in either case, gravity can only serve to pull things together. It can't *push*; and a push is what you need in order to explain how the expansion of the universe can possibly be accelerating.

Where could such a push—a repulsive force between the galaxies—possibly come from? The only possible source, according to today's science, arises from space itself. There must be a "vacuum energy," present even in empty space, and providing an expansion force powerful enough to overcome the attraction of gravity. The idea of such a source of force was introduced by Einstein over eighty years ago, as a so-called "cosmological constant." Einstein used this constant to explain why the universe did *not* expand (this was before Hubble's observations showing that it did) and Einstein called his failure to imagine an expanding universe the biggest blunder of his life. Until recently, most cosmologists preferred to assume that the value of the cosmological constant was zero, which meant there was no repulsive force associated with space itself.

The new observations of an accelerating universe imply that this is no longer an option. The cosmological constant can't be equal to zero if space itself is to be the origin of a repulsive force more than strong enough to balance gravitational attraction.

And now for the *so what?*: Can such esoteric ideas, originating so far away, have any relevance to everyday life?

I can't really answer that. But I will point out that

the proposed vacuum energy is present here on Earth, as well as in remote locations. And notions equally abstract, published by Einstein in 1905 and concerning the nature of space and time, led very directly to atomic energy and the atomic bomb. That development took less than forty years. If history is any guide, many of us might live to see practical consequences of a non-zero cosmological constant.

A.12. Nothing but blue skies . . . Let me describe a condition: it is a physical disability that affects more than twenty million Americans; it is usually congenital, and almost always incurable; it is at best a nuisance, and it is at worst life-threatening.

You might think that such an ailment would be a major item on the agenda of the National Institutes of Health, perhaps even the subject of a Presidential Commission to seek urgent action.

No such thing. The condition I have described is color blindness. It is strongly sex-linked. One man in every twelve suffers from it to some extent, compared with only one woman in two hundred. And the whole subject enjoys little attention.

Part of the reason for our lack of emphasis on color blindness is its invisibility. You can't tell that a man has such a disability, though his choice of shirt and matching tie may be a bit of a giveaway. In fact, you may suffer some form of it yourself and become aware of that fact only in special circumstances. In my own case I have difficulty distinguishing blues and greens, but I only notice it when playing "Trivial Pursuit"; then I am never sure if I have a blue or a green question coming.

A more common—and a more dangerous—form cannot distinguish red from green. John Dalton, the chemist, a colorblind person and one of the first people to write about it, reported that "blood looks like bottle-green and a laurel leaf is a good match for sealing wax." In more modern times, sufferers are forced to distinguish the condition of traffic lights by their vertical placement, and they are at risk in situations where a red "Stop" light or a green "Go" light offers no other information to back it up.

The problem originates in the retina, at the back of our eyes. The retina contains two different kinds of light-sensitive objects, each microscopic in size. The retinal *rods* do not perceive color at all, and they are most useful at low light levels. The retinal *cones* are responsible for all color vision, but they need a higher light level before they become sensitive. Recall that, on a moonlit walk, your surroundings are rendered only in black and white.

In a person with normal vision, the signals generated by the cones and transmitted by nerve cells to the brain permit all color to be distinguished. If you are color blind, however, certain colors will produce the same signals as each other. Green and red may be confused, or pale green and yellow, or, in my case, certain greens and blues.

Unlike some other conditions, color blindness has few compensating advantages. In a military situation, a color blind person may detect camouflage which fools ordinary eyes, but in general, color blindness is nothing but a nuisance. So can we do anything about it?

As I said at the beginning, this disability is usually incurable. It can, however, often be alleviated by the use of special eyeglasses. These contain filters that modify the light passing through them, in such a way that they convert different colors to combinations that the wearer can perceive.

That's today, and even if impressive it's pretty crude. Twenty years from now we will be able to go much farther, with personalized "false color" eyeglasses. These will employ sensors and displays that transform any light falling on them into color regions for which the wearer's retina is able to generate distinguishable signals.

You might argue that a person with such eyeglasses is still color blind, because his perception of blue, green, or red will be different from yours. If you make this point, I will ask the question: how do you know that what you perceive when you see colors is at all the same as other people's?

A.13. Thinking small. The launch of a space shuttle is an impressive event. It is impressively big, impressively noisy, and impressively expensive. During the first

few minutes of ascent, energy is used at a rate enough
to power the whole United States. Most of us love fireworks, and I have never met anyone who did not enjoy
watching this fireworks display on the grandest scale.

On the other hand, is this a *necessary* display of size
and power? We are in the habit of thinking that sending something into space requires a vast and powerful
rocket, but could we be wrong?

We could, and we are. Most of our preconceived ideas
about rocket launches go back to the early days of the
"space race" between the United States and the Soviet
Union, and in the 1950s and 1960s the name of the game
was placing humans into orbit. There were good psy-
chological and practical reasons for wanting to send
astronauts and cosmonauts. First, the public is always
far more interested in men than in machines; and second, the computers of the early days of the space program were big, primitive and limited in what they could
do. People, by contrast, possessed—and possess—far more
versatility than any computer, and can perform an endless
variety of tasks.

On the other hand, people come in more-or-less standard sizes. They also need to eat, drink and breathe.
Once you decide that humans are necessary in space,
you have no alternative to big rockets; but today's
applications satellites, for communications, weather
observation, and resource mapping, neither need nor want
a human presence.

As computers become smarter and more powerful, they
are also shrinking in size and weight. The personal
computer in your home today is faster and has far more
storage than anything in the Apollo program spacecraft.
Other electronics, for observing instruments and for
returning data to the ground, is becoming micro-
miniaturized. Payloads can weigh less. So how big—or
how small—can a useful rocket be?

We are in the process of finding out. Miniature thrust
chambers for rocket engines have already been built, each
one smaller and lighter than a dime. A group of about
a hundred of these should be able to launch into orbit
something about the size of a Coke can. That's more

than big enough to house a powerful computer, plus an array of instruments. One of these "microsats" could well become an earth resources or weather observing station.

We are at the very beginning of thinking small in space. How small might we go? Since our experience with space vehicles is limited, let us draw an analogy with aircraft. Today's aircraft, like today's spacecraft, are designed to carry people. Suppose, however, that we just want a flying machine that can carry a small payload (maybe a few grams, enough for a powerful computer). How small and light can it be? We don't have a final answer to that question, although today a jet engine the size of a shirt button is being built at MIT. However, Nature provides us with an upper limit on size. Swallows, weighing just a few ounces, every year migrate thousands of miles without refueling. We should be able to do at least this well.

And if you want to think really small, look at what the swallows eat: flying insects, each with its own on-board navigation and observing instruments. Imagine a swarm of space midges, all launched on a rocket no bigger than a waste paper basket, each one observing the Earth or the sky and returning their coordinated observations back to the ground. Imagination could become reality in less than half a century.

A.14. New maps for old. Map-making in ancient times was not a job for the faint-hearted.

Even without the early worries of going too far and sailing off the edge of the world, anyone interested in determining the positions of land masses and shore lines had to face the dangers of reefs, shoals, storms at sea, scurvy, shipwreck, and starvation. Perhaps even worse were the hostile natives met along the way, who killed, among others, the famous explorers Ferdinand Magellan and Captain James Cook.

Mapping the interior of a country was just as difficult. The hardy surveyor had to face deserts, glaciers, avalanches, impassable rivers, infectious diseases, dangerous animals, and still more hostile natives.

And yet maps were early recognized as vitally important. Within settled countries they were needed to define property ownership, set taxes, measure land use, and establish national boundaries. Farther afield, the lack of good maps and accurate knowledge of position led to countless shipwrecks. In 1707, an English fleet commanded by the splendidly-named Sir Cloudesley Shovel made an error in navigation, ran ashore on the Scilly Isles which they thought were many miles away, and lost more than two thousand sailors.

Why was map-making so hard? It sounds easy. All you need to define a point on the surface of the Earth uniquely are three numbers: latitude, longitude, and height above some reference surface (usually sea-level). Measure a few thousand or tens of thousands of such points, and you have an accurate map of the Earth.

Unfortunately, it was difficult verging on impossible to determine absolute locations. The fall-back position was to measure *relative* locations. Starting with a baseline a few miles long, a distant point was identified, and accurate angles from each end of the baseline were measured. This allowed the other two sides of the triangle to be calculated; from these as new baselines, new angle measurements led to more triangles, which led to still more triangles, until finally the whole country or region was covered by a network. In practice, because there could be small errors in each measurement, all the angles and lengths in the network were adjusted together to produce the most consistent result.

What we describe sounds straightforward, but the amount of measurement and computation in a large mapping survey was huge. The calculations were, of course, all done by hand. A survey of this type could take years, or even decades. There was also no substitute for going out and making ground measurements. Even fifty years ago, it was possible for a leading expert on maps to declare, with perfect confidence, "there is only one way to compile an accurate map of the earth . . . and that is to go into the field and survey it."

Today, that is not the case at all. The new generation of map-makers sit in their offices, while far above them, satellites look down on the Earth and send back a continuous stream of images revealing details as small as a few feet across. In perennially cloudy regions, spaceborne radar systems see through to the ground below. The location of the images is known fairly well, but not accurately enough to make good maps. However, the images can be cross-referenced, by identifying common ground features on neighboring and overlapping images. Also, the position of selected points on the ground can be found absolutely, to within a few tens of meters, using another satellite system known as GPS (the Global Positioning System).

Finally, all the image data and all the cross-reference data can be adjusted simultaneously, in a computer calculation of a size that would have made all early map-makers blench. The result is not just a map of the Earth—it is an *accurate* map and a *recent* map, in which a date can be assigned to any observed feature.

As the people involved in this will tell you, it is still hard work—but it sure beats cannibals and shipwrecks.

A.15. The ears have it. I am one of those unfortunate people who have trouble singing the "Star-Spangled Banner." It's not that I don't know the tune, it's that my useful vocal range is only about one octave. The National Anthem spans an octave and a half. No matter where I start with "Oh say can you see," by the time I get to "the rockets' red glare" I sound like a wolf baying at the moon.

I comfort myself with the thought that humans are primarily visual animals. Eighty percent, maybe even ninety percent, of the information that we receive about the world comes to us as visual inputs. Bats, by comparison, depend mainly on sound, "seeing" the world by echolocation of reflected sound signals that they themselves generate. And as for the other senses, any dog owner will tell you that an object without a smell counts as little or nothing in the canine world.

Being human, we have a tendency to argue for the superiority of "our" primary way of perceiving the world. After all, we have stereoscopic, high-definition, full color vision, and that's a rare ability in the animal kingdom.

But would an intelligent bat agree with us, or would it be able to make a good case for its own superior form of perception?

Let's compare sound and light. They may seem totally different, but they have many similarities. Both travel as wave forms, and both can be resolved into waves of different single frequencies (colors, in the case of light). The note that we hear as middle C has a wavelength of a little more than four feet, whereas what we see as the color yellow has a wavelength of only one twenty-millionth of that. Also, sound waves need something— air, water, metal—to travel through, while light waves travel perfectly well through a vacuum. No bat can ever see the stars. However, I would argue that these are unimportant differences. We have equipment that can readily translate sounds to light, or convert different colored light to sounds.

Our intelligent bats would agree with all of this; but what they would point out, quite correctly, is that our visual senses lack *range*. We can hear, with no difficulty, sounds that go all the way from thirty cycles a second, the lowest note on a big pipe organ, to fifteen thousand cycles a second, beyond the highest note of the piccolo. That is a span of nine octaves (an octave is just the doubling of the frequency of a note). Compare this with our eyesight. The longest wavelength of visible light (dark red) is not quite twice the wavelength of the shortest light that our eyes can detect (violet). The range of what we can see is less than one octave. If we were to convert "The Star-Spangled Banner" to equivalent light, not a person on earth would be able to see the whole thing.

Why can we observe such a limited range of wavelengths, while hearing over a vastly greater one? It is a simple matter of the economy of nature. Our eyes have adapted over hundreds of millions of years to be sensitive in just the wavelength region where the sun produces its maximum illumination. The amount of radiation coming from the sun falls off rapidly in the infrared, at wavelengths longer than what we can see, while waves much shorter than violet are absorbed strongly by the atmosphere (lucky for us, or we would fry).

Of course, being the inventive monkeys that we are, humans have found ways around the natural limitations of our eyes. Today we have equipment that provides pictures using everything from ultra-short X-rays to mile-long radio waves. We roam the universe, from the farthest reaches of space to the insides of our own bodies. With the help of our instruments, we can observe not just nine or ten octaves, but more than forty. Let's see the bats match that one.

A.16. Memories are made of—what? Over the years I have met many people in many professions: actors, writers, biologists, computer pioneers, artists, astronomers, composers, even a trio of Nobel Prize winners in physics. They had numerous and diverse skills. What none of them had was a good memory. Or rather, what none of them would *admit to* was a good memory. Their emphasis was the other way round: how hard it was to recall people's faces, or names, or birthdays, or travel directions.

History records examples of people with prodigious memories. Mozart, at thirteen, went to the Sistine Chapel in Rome to hear a famous *Miserere* by Allegri, then wrote out the whole work. The mathematician, Gauss, did not need to look up values in logarithm tables, because he knew those tables by heart. And Thomas Babington, Lord Macaulay, seemed to have read so much and remembered it so exactly that one of his exasperated colleagues, Lord Melbourne, said, "I wish I was as cocksure of *anything* as Tom Macaulay is of everything."

To the rest of us, hard-pressed to remember our own sister's phone number, such monster memories seem almost inhuman. My bet, however, is that even these people would, if asked, complain of their poor memories and emphasize what they forgot. And each of us, without ever thinking about it, has enormous amounts of learned information stored away in our brain.

I say "learned information," because some of what we know is *hard-wired*, and we call that instinct. We don't learn to suck, to crawl, or to walk by committing actions to memory, and we normally reserve the word "memorize" to things that we learn about the world through observation and experience. I am going to stick with this distinction between instinct and memory, though sometimes the borderline becomes blurred. We don't remember learning to talk, but we accept that it relies on memory because others tell us we did (though there is good evidence that the *ability* to acquire language is hardwired). And most of us would not say that riding a bicycle depends on memory, although clearly this is a learned and not an inborn activity.

I want to concentrate on factual information that is definitely learned, stored, and recalled, and ask two simple questions: Where is it stored, and how is it stored?

The easy part first: information is stored in the brain. But when we ask *where* in the brain, and ask for the form of storage, we run at once into problems. The tempting answer, that a piece of data is stored in a single definite location, as it would be in a computer, proves to be wrong. Although many people believe that the brain ultimately operates like a computer—a "computer made of meat"—in this case the analogy is more misleading than helpful.

Much of what we know about memory comes from the study of unfortunate individuals with brains damaged by accident or disease. This is hardly surprising, since volunteers for brain experiments are hard to come by (as Woody Allen remarked, "Not my brain. It's my second favorite organ."). Studies of abnormal brains can be misleading, but they show unambiguously that a human memory does not sit in a single defined place. Rather, each memory seems to be stored in a distributed form, scattered somehow in bits and pieces at many different physical locations. Although ultimately the information must be stored in the brain's neurons (we know of nowhere else that it *could* be stored), we do not yet understand the mechanism. Some unknown process hears the question, "Who delivered the Gettysburg address?", goes off into the interior of the brain, finds and assembles information, and returns the answer (or occasionally, and frustratingly, fails to return the answer): "Abraham Lincoln."

And it does the job *fast*. The brain contains a hundred billion neurons, but the whole process, from hearing the question to retrieving and speaking the answer, takes only a fraction of a second.

We may not be Mozart, but each of us possesses an incredible ability to store and recall information. And are we impressed by this? Not at all. Instead of being pleased by such a colossal capability, we are like the celebrated Mr. X, always complaining about his sieve-like memory.

I would give Mr. X's name, but at the moment I cannot quite recall it.

A.17. In defense of Chicken Little. Chicken Little wasn't completely wrong. Some of the sky does fall, some of the time. When a grit-sized particle traveling at many miles a second streaks into the Earth's atmosphere and burns up from friction with the air before reaching the ground, we call it a shooting star or a meteor. Some of us make a wish on it. We think of meteors as harmless and beautiful, especially when they come in large groups and provide spectacular displays such as the Leonid and Perseid meteor showers.

Meteors, however, have big brothers. These exist in all sizes from pebbles to basketballs to space-traveling mountains. If the speeding rock is large enough, it can remain intact all the way to the ground and it is then known as a meteorite. The reality of meteorites was denied for a long time—Thomas Jefferson said, "I could more easily believe that two Yankee professors would lie than that stones would fall from heaven"—but today the evidence is beyond dispute.

If one of these falling rocks is big enough, its great speed gives it a vast amount of energy, all of which is released on impact with the Earth. Even a modest-sized meteorite, twenty meters across, can do as much damage as a one-megaton hydrogen bomb. This sounds alarming, so let us ask three questions: How many rocks this size or larger are flying around in orbits that could bring them into collision with the Earth? How often can impact by a rock of any particular size be expected? And how does damage done vary with the size of the meteorite?

Direct evidence of past impacts with Earth is available only for large meteorites. For small ones, natural weathering by wind, air, and water erases the evidence in a few years or centuries. However, we know that a meteorite, maybe two hundred meters across, hit a remote region of Siberia called Tunguska, on June 30, 1908. It flattened a thousand square kilometers of forest and put enough dust into the atmosphere to provide colorful sunsets half a continent away. About 20,000 years ago, a much bigger impact created Meteor Crater in Arizona, more than a kilometer across. And 65 million years ago, a monster meteorite, maybe ten kilometers across, struck in the Gulf of Mexico. It caused global effects on weather, and is believed to have led to the demise of the dinosaurs and the largest land reptiles.

The danger of impact is real, and beyond argument. But is it big enough for us to worry about? After all, sixty-five million years is an awfully long time. How do we make an estimate of impact frequency?

The answer may seem odd: we look at the Moon. The Moon is close to us in space, and hit by roughly the same meteorite mix. However, the Moon is airless, waterless, and almost unchanging, so the history of impacts there can be discovered by counting craters of different sizes. Combining this with other evidence about the general size of objects in orbits likely to collide with Earth, we can calculate numbers for frequency and energy release. They are not totally accurate, but they are probably off by no more than a factor of three or four.

I will summarize the results by size of body, and translate that to the equivalent energy released as number of megatons of H-bombs. About once a century, a "small" space boulder about five meters across will hit us and produce a matching "small" energy equal to that released by the Hiroshima atomic bomb. It will probably burn up in the atmosphere and never reach the ground, but the energy release will be no less. Once every two thousand years, on average, we will get hit by a twenty-meter boulder, with effects a little bigger than a one-megaton H-bomb. Every two million years, a five-hundred-meter giant will arrive, delivering as much energy as a full-scale nuclear war.

I found these numbers disturbing, so a few years ago I sent them to the late Gene Shoemaker, an expert on the bombardment of Earth by rocks from space. He replied, not reassuringly, that he thought my numbers were in the right ballpark, but too optimistic. We will be hit rather more often than I have said.

Even if I were exactly right, that leaves plenty of room for worry. Being hit "on average" every 100,000 years is all very well, but that's just a *statistical* statement. A big impact could happen any time. If one did, we would have no way to predict it, or—despite what recent movies would have you believe—prevent it.

A.18. Language problems and the Theory of Everything.

A couple of weeks ago I received a letter in Spanish. I don't know Spanish. I was staring at the text, trying and failing to make sense of it by using my primitive French, when my teenage daughter wandered by. She picked up my letter and cockily gave me a quick translation.

I was both pleased and annoyed—aren't I supposed to know more than my children?—but the experience started me thinking: about language, and the importance of the right language if you want to do science in general, and physics in particular.

Of course, every science has its own special vocabulary, but so does every other subject you care to mention. Partly it's for convenience, although sometimes I suspect it's a form of job security. Phrases like "stillicide rights," "otitis mycotica," and "demultiplexer" all have perfectly good English equivalents, but they also serve to sort out the insiders from the outsiders.

One subject, though, is more like an entire language than a special vocabulary, and we lack good English equivalents for almost all its significant statements. I am referring to mathematics; and, like it or not, modern physics depends so heavily on mathematics that non-mathematical versions of the subject mean very little. To work in physics today, you have to know the language of mathematics, and the appropriate math vocabulary and methods must already exist.

On the face of it, you might think this would make physics an impossibly difficult subject. What happens if you are studying some aspect of the universe, and the piece of mathematical language that you need for its description has not yet been invented? In that case you will be out of luck. But oddly—almost uncannily—throughout history, the mathematics had already been discovered before it was needed in physics.

For example, in the seventeenth century Kepler wanted to show that planets revolved around the Sun not in perfect circles, but in other more complex geometrical figures. No problem. The Greeks, fifteen hundred years earlier, had proved hundreds of results about conic sections, including everything Kepler needed to know about the ellipses in which planets move. Two hundred years later, Maxwell wanted to translate Michael Faraday's experiments into a formal theory. The necessary mathematics, of partial differential equations, was sitting there waiting for him. And, to give one more example, when Einstein's theory of general relativity needed a precise way to describe the properties of curved space, the right mathematics had been created by Riemann and others and was already in the text books.

Of course, there can be no guarantee that the mathematical tools and language you want will be there when you need it. And that brings me to the central point of this column. One of the hottest subjects in physics today is the "Theory Of Everything," or TOE. The "Everything" promised here is highly limited. It won't tell you how a flower grows, or explain the IRS tax codes. But a TOE, if successful, will pull together all the known basic forces of physics into one integrated set of equations.

Now for the tricky bit. The most promising efforts to create a TOE involve something known as string theory, and they call for a description of space and time far more complicated than the height-width-length-time we find adequate for most purposes. The associated mathematics is fiendishly difficult, and is not just sitting in the reference books waiting to be applied. New tools are being created, by the same people doing the physics, and it is quite likely that these will prove inadequate. The answers may just have to wait, until, ten or fifty years from now, the right mathematical language has been evolved and can be applied.

It's one of my minor personal nightmares. Mathematics, more than almost any other subject, is a game played best by the young. Suppose that, five or fifteen years from now, we have a TOE that explains everything from quarks to quasars in a single consistent set of equations. It will, almost certainly, require for its understanding some new mathematical language. By that time I may just be too old or set in my ways ever to learn what's needed.

It's a dismal prospect. You wait your whole life for something, and then when it finally comes along you find you can't understand it.

A.19. Fellow travelers. My mother grew up in a household with nine children and little money. Not much was wasted. Drop a piece of food on the floor and you picked it up, dusted it off, and ate it. This doesn't seem to have done my mother much harm, since she is still around at ninety-seven. Her philosophy toward food and life can be summed up in her comment, "You eat a peck of dirt before you die."

Contrast this with the television claim I heard a couple of weeks ago: "Use this product regularly, and you will rid yourself and your house completely of germs and pests."

The term "pest" was not described. It probably didn't include your children's friends. But whatever the definition, the advertisers are kidding themselves and the public by making such extravagant claims. Your house, and you yourself, are swarming with small organisms, whose entry to either place was not invited but whose banishing is a total impossibility.

I have nothing against cleanliness, and certainly no one wants to encourage the presence in your home of the micro-organisms that cause cholera, malaria, bubonic plague, and other infectious diseases. Such dangers are, however, very much in the minority. Fatal diseases are also the failures among the household invaders. What's the point of invading a country, if the invasion makes the land uninhabitable? In our case, that amounts to the organism infecting and killing its host. Successful invaders don't kill you, or even make you sick. The most successful ones become so important to you that you could not live without them.

Biologists set up a hierarchy of three types of relationship between living organisms. When one organism does nothing but harm to its host, that's called *parasitism*. In our case, this includes things like ringworm, pinworms, athlete's foot, ticks, and fleas. All these have become rarer in today's civilized nations, but most parents with children in elementary school have heard the dread words "head lice," and have probably dealt with at least one encounter.

Parasites we can do without. This includes everything from the influenza virus, far too small to see, to the tapeworm that can grow to twenty feet and more inside your small intestine.

Much more common, however, are the creatures that live on and in us and do neither harm nor good. This type of relationship is known to biologists as *commensalism*. We provide a comfortable home for tiny mites that live in our eyelashes, to others that dine upon cast-off skin fragments, and to a wide variety of bacteria. We are unaware of their presence, and we would have great difficulty ridding ourselves of them. It might even be a bad idea, since we can't be sure that they do not serve some useful function.

And then there is *symbiosis*, where we and our fellow-traveling organisms are positively good for each other. What would happen if you could rid yourself of all organisms that do not possess the human genetic code?

The answer is simple. You would die, instantly. In every cell of your body are tiny objects called mitochondria. They are responsible for all energy generation, and they are absolutely essential to your continued existence. But they have their own genetic material and they reproduce independently of normal cell reproduction. They are believed to be bacteria, once separate organisms, that long ago entered a symbiotic relationship with humans (and also with every other animal on earth).

If the absence of mitochondria didn't kill you in a heartbeat, you would still die in days. We depend on symbiotic bacteria to help digest our food. Without them, the digestive system would not function and we would starve to death.

"We are not alone." More and more, we realize the truth of that statement. We are covered on the outside and riddled on the inside by hundreds of different kinds of living organisms, and we do not yet understand the way that we all relate to each other. For each, we have to ask, is this parasitism, commensalism, or symbiosis?

Sometimes, the answers are surprising. Twenty years ago, gastric ulcers were blamed on diet or stress. Today, we know that the main cause is the presence in the stomach of a particular bacterium known as *Helicobacter pylori*. Another organism, *Chlamydia*, is a suspect for coronary disease and hardening of the arteries. A variety of auto-immune diseases may be related to bacterial action.

All these facts encourage a new approach for biologists and physicians: The best way to study humans is not as some pure and isolated life form; rather, each of us should be regarded as a "superorganism." The life-cycles and reproductive patterns of us and all our fellow travellers should be regarded as one big interacting system.

Disgusting, to be lumped in with fleas and mites and digestive bacteria, as a single composite object? I don't think so. In a way it's a comforting thought. We are not alone, and we never will be.

A.20. How do we know what we know? At the moment there is a huge argument going on about the cause of AIDS. Most people in this country—but by no means all—believe that the disease is caused by a virus known as HIV, the Human Immunodeficiency Virus. In Africa, however, heads of governments have flatly stated that they don't accept this. They blame a variety of other factors, from diet to climate to genetic disposition.

The available scientific evidence ought to be the same

for everyone. So how can there be such vast differences in what people believe?

Part of the reason is what we might call the "Clever Hans" effect. Clever Hans was a horse who lived in Germany early in the twentieth century, and he seemed to be smarter than many of the humans around him. He could answer arithmetic problems by tapping out the correct answers with a fore-hoof, and give yes or no answers to other questions—Is London the capital of France?—by shaking or nodding his head, just like a human.

His owner, a respected Berliner named Wilhelm von Osten, was as astonished as anyone by Clever Hans' abilities. There seemed no way that he would commit fraud, particularly since Clever Hans could often provide correct answers when von Osten was out of the room, or even in a different town. The Prussian Academy of Sciences sent an investigating committee, and they too were at first amazed by the horse's powers. True, there were inconsistencies in the level of performance, but those could often be explained away.

Finally, almost reluctantly, the truth was discovered. Clever Hans could not do arithmetic, and did not know geography and history. He was responding to the body language of the audience. Most observers, including members of the investigating committee, wanted Hans to get the right answers. So they would instinctively tense at the question, and relax when Hans gave the right answer. The body movements were very subtle, but not too subtle for Hans. He really was clever—clever at reading non-verbal cues from the humans around him.

We are no different from the groups who met Clever Hans. We all want certain answers to be true. Given a mass of evidence, we tend to notice the facts that agree with our preferences, while explaining away the inconvenient ones that would tell us otherwise. And AIDS is a disease so complex and so widespread that you can find what appear to be exceptions to any general rules about its cause, spread, or inevitable effects.

That, however, is only half the story. The other reason there can be such intense arguments about AIDS applies equally well to half the things—or maybe today it's ninety-nine percent of the things—in our lives. We have actual experience in certain areas: boiling water hurts; you can jump off a ten-foot ladder but you can't jump back up; the moon will be full about once a month; it's colder in winter than in summer; coffee with salt instead of sugar tastes terrible.

But there are a million other things in everyday life for which we have no direct experience and explanation. Can you tell me how a digital watch works? Why is a tetanus shot effective for ten years, while even with an annual flu shot you are still likely to get the flu? What does that computer of yours do when you switch it on? How does e-mail from your computer travel across the country to a friend on the opposite coast, or halfway around the world? Just what is plastic, and how is it made? How does your refrigerator work? When you flip a light switch, where does the electricity come from? It's not like turning on a faucet, where we know that somewhere a huge reservoir of water sits waiting to be tapped. So how come the electricity is there just when you need it?

I can give answers to these, in a hand-waving sort of fashion, but if I want any sort of details I have to go and ask questions of specialists whom I trust. And most of the questions that I've just asked are not new, or even close to new. The refrigerator was patented in 1834. The first plastics, like our electricity supply, go back to the beginning of the twentieth century.

Good answers are available to every one of my questions, all we have to do is seek them out. But what about the newer areas of research, for which AIDS forms a fine example? When the experts themselves are still groping their way toward understanding, and still disagreeing with each other, what chance do the rest of us have?

Not much, provided that we insist on direct evidence. Every one of us must decide for ourselves who and what to believe. We, like the audience of Clever Hans, are going to believe what we want to believe until evidence to the contrary becomes awfully strong.

And maybe even after that. We, as ornery humans, tend to go on believing what we prefer to believe.

A.21. Where are they? Our "local" galaxy contains about a hundred billion stars. We see only a few thousand of the closest as actual points of light, though millions of others merge into a broad and diffuse glow that we notice on clear nights and call the Milky Way.

A hundred billion is such a big number that it's hard to have a real feel for it, so let's put it this way: there are enough stars in our galaxy for every human on earth to own sixteen apiece. Not only that, our galaxy is just one of the hundred billion galaxies that make up the known universe. If humans owned the whole cosmos, each of us could lay claim to more than a trillion stars. That's the astronomical equivalent of everyone being owed the National Debt, with each star and its planets priced at about a dollar.

Of course, there's a big "if" in there. We can only claim the universe if no others are out there to stake counterclaims and assert property rights. Which leads to the big question: Are there other living beings in the universe, at least as intelligent as we are; or are we the only smart, self-aware objects in creation? As the late Walt Kelly remarked, long ago, either way it's a mighty sobering thought.

Some people insist that intelligent aliens in the universe have appeared right here on Planet Earth, occasionally taking selected individuals for a space ride but otherwise keeping a low profile. I am not in that group of believers. I can't see why anyone would bother to travel such gigantic distances and then remain in hiding. The idea that aliens have actually crash-landed in remote parts of the country, and had their presence covered up by the government, has even less appeal. If anywhere, Washington, D.C., is the place to look for aliens.

Let's take another approach. We have a rough idea of the total number of stars in all the galaxies. How many of those stars have planets? Ten years ago we had no direct evidence of any, but today some new planet around another star is discovered at least once a month. Suppose, then, that only one star in a thousand has a planet around it—a very low estimate. That still gives us a hundred million planets as candidates right here in our own galaxy. If just one percent of those can support life, a million other worlds have living things on them.

The next step is the hardest one. If a world has life, what are the chances that one of those living creatures will develop intelligence and technology, enough to build a starship, or at least to send out a signal to us?

We don't know. Let me state that more strongly: we have not the slightest idea. But we can listen, and we do, for evidence of alien existence. We listen not with sound waves, but with radio waves. For the past forty years, a search for extraterrestrial intelligence (SETI) program has been carried on in this country and around the world. Using radio telescopes capable of picking up the tiniest trickle of energy, we eavesdrop on the sky and hope to discover the organized series of pulses that would announce the presence of other thinking beings.

So far we have found nothing. This is sometimes called the Great Silence, sometimes the Fermi Paradox (Fermi asked the simple question, "Where are they?"). On the other hand, forty years of listening is no time at all in a universe at least ten billion years old, particularly since the SETI program is run on a shoestring. It has no government funding. It is paid for and operated by people who believe that a positive result to the search would change the way we think about everything.

Speaking for myself, I would just love to change the way we think. For instance, if we were willing to spend as much money listening to the stars as we do on, say, land mines, we might detect and decipher that world-altering message from the sky.

Are we alone in this galaxy, as an intelligent life form? It is hard to imagine a more profound question. I'd gladly give up any claim to the trillion-plus stars that represent my share of the universe, to know the answer.

And while I'm at it, I'll gladly give up my share of land mines.

Note: If you own a personal computer and a modem, you can become directly involved in analyzing radio data that may contain evidence of alien signals. Contact me if you would like to know how.

REFERENCES

Adleman, Leonard, "Molecular Computation of Solutions of Combinatorial Problems"; *Science*, Volume 266, 1994.

Anderson, Philip, "More is Different," *Science*, 1972.

Anderson, Poul, *Tau Zero*; 1970.

Arrhenius, Svante, *Worlds in the Making*; 1908 (translated from the German, *Werden Der Welten*, 1907).

Baxter, Stephen, *Flux*; 1993, HarperCollins (U.S. printing 1993).

Bear, Greg, "Blood Music," *Analog*, 1983; see also the novel *Blood Music*; Ace Books, 1985.

Benford, Gregory, *Timescape*; Simon & Schuster, 1979.

Benford, Gregory, and William Rotsler, *Shiva Descending*; Avon, 1980.

Bergmann, Peter; *The Riddle of Gravitation*; Scribners, 1968.

Bester, Alfred, *The Demolished Man*; Shasta, 1953.

——— *The Stars My Destination*; Signet, 1956.

Bondi, Hermann, *Relativity and Common Sense*; Dover 1981 reprint of Doubleday original volume, 1964.

Born, Max, *Einstein's Theory of Relativity*; Dover 1965 reprint of Methuen original volume, 1924.

Bradbury, Ray, "A Sound of Thunder"; Collier's, 1952.

Brin, David, *Earth*; Bantam Books, 1990.

Cairns-Smith, A. G., *Genetic Takeover and the Mineral Origins of Life*; Cambridge University Press, 1982.

———— *The Life Puzzle*; Cambridge University Press, 1971.

———— *Seven Clues to the Origin of Life*; Cambridge University Press, 1985.

Calder, Nigel, *Einstein's Universe*; Greenwich House, 1979.

Capek, Karel, *R.U.R.: A Fantastic Melodrama*; 1920. (Capek's robots would today be called androids.)

Clarke, Arthur C., *The Fountains of Paradise*; Del Rey, 1979.

———— *Imperial Earth*; Del Rey, 1975.

———— *The Songs of Distant Earth*; Del Rey, 1986.

Clement, Hal, *Close to Critical*; Ballantine, 1964.

———— *Cycle of Fire*; Ballantine, 1957.

———— *Iceworld*; Gnome, 1953

———— *Mission of Gravity*; Doubleday, 1953.

Dawkins, Richard, *Climbing Mount Improbable*; W.W. Norton and Co., 1996.

———— *River Out of Eden*; Basic Books, 1995.

———— *The Selfish Gene*; Oxford University Press, 1976.

Drexler, K. E., *The Engines of Creation*; Viking, 1986.

———— (with Chris Peterson and Gayle Pergamit), *Unbounding the Future*; William Morrow, 1991.

Dyson, Freeman, *Disturbing the Universe*; Basic Books, 1979.

———— "Time Without End: Physics and Biology in an Open Universe"; *Reviews of Modern Physics*, Volume 51, 1979.

Eccles, John, *How the Self Controls Its Brain*; Springer-Verlag, 1994.

Einstein, Albert, *The Meaning of Relativity*; Easton Press, 1994; reprint of original Princeton University Press volume, 1953.

Forward, Robert L., *Dragon's Egg*; Del Rey, 1980.

———— *Future Magic*; Avon Books, 1988.

———— *Rocheworld*; Baen Books, 1990; an expansion of the novel *The Flight of the Dragonfly*; Timescape Books, 1984.

Fracastoro, Girolamo, *De Contagione*; 1546.

Furusawa, A., J. L. Sorensen, S. L. Braunstein. C. A. Fuchs, H. J. Kimble, and E. S. Polzik, "Unconditional Quantum Teleportation"; *Science,* Volume 282, pages 706–709, October 23, 1998.

Gleick, James, *Chaos: Making a New Science*; Viking, 1987.

Godwin, Francis, *The Man in the Moone*; 1638.

Gribbin, John, *In Search of Schrödinger's Cat*; Bantam, 1984.

Haldane, J. B .S., "On Being the Right Size," essay originally published in 1927; reprinted in the collection *On Being the Right Size and Other Essays*, John Maynard Smith, Editor; Oxford University Press, 1985.

Heinlein, Robert, *The Puppet Masters*; Doubleday, 1951.

Henderson, Zenna, *The People: No Different Flesh*; Doubleday, 1966.

——— *Pilgrimage*; Doubleday, 1961.

Herbert, Frank, *Dune*; Chilton, 1965.

Hogan, James P., *Paths to Otherwhere*; Baen Books, 1996.

Horgan, John, *The End of Science: Facing the Limits of Knowledge and the Twilight of the Scientific Age*; Addison-Wesley, 1996.

Hoyle, Fred, and Chandra Wickramasinghe, *Diseases from Space*; Sphere Books, 1977.

Hutton, James, *Theory of the Earth*; 1795.

Huxley, Aldous, *After Many a Summer Dies the Swan*; Harper, 1939.

Kepler, Johannes, *Somnium*; 1634.

Kingsbury, Donald, "To Bring in the Steel"; *Analog*, 1978.

Kippenhahn, R., *100 Billion Suns*; Basic Books, 1983. Originally published in German in 1980.

Kress, Nancy, "Feigenbaum Number"; *Omni*, 1995.

——— *Maximum Light*; Tor Books, 1998.

——— *Oaths and Miracles*; Tor Books, 1996.

Lindley, David, *The End of Physics: The Myth of a Unified Theory*; Basic Books, 1993.

Lipton, Richard J., "DNA Solution of Hard Computational Problems"; *Science*, Volume 268, 1995.

Lovelock, James E., *Gaia: A New Look at Life on Earth*; Oxford University Press, 1979.

Mayr, Ernst, *The Growth of Biological Thought*; Harvard University Press, 1982.

McDevitt, Jack, *The Hercules Text*; 1986.

——— *Moonfall*; HarperPrism, 1998.

McDonough, Thomas R., *The Search for Extraterrestrial*

Intelligence: Listening for Life in the Cosmos; John Wiley and Sons, 1987.

Minsky, Marvin, *The Society of Mind*; Simon and Schuster, 1986.

Moravec, Hans, *Mind Children: The Future of Robot and Human Intelligence*; Harvard University Press, 1988.

Morrison, Philip, and Giuseppe Cocconi, *Nature*; 1959.

Morrison, Philip, J. Billingham, and J. Wolfe, editors, *The Search for Extraterrestrial Intelligence: SETI*; NASA Publication SP-419, 1977.

Niven, Larry, "The Hole Man"; *Analog*, 1973; reprinted in *A Hole in Space*; Ballantine, 1974.

——— *Ringworld*; 1970.

——— Tales of Known Space, includes the separate volumes *Ringworld*, *The Ringworld Engineers*, *The Ringworld Throne*, *Protector*, *World of Ptaavs*, and *A Gift from Earth*, as well as numerous short stories.

Niven, Larry, and Jerry Pournelle, *Lucifer's Hammer*; Playboy Press, 1977.

Nordley, G. David, "Into the Miranda Rift"; *Analog*, 1993.

Oppenheimer, J. R., and H. Snyder, "On continued gravitational contraction"; *Physical Review*, Volume 56, pages 455–459, 1939.

Pauli, Wolfgang, *Theory of Relativity*; Dover 1981 reprint of the original 1921 monograph, with Pauli's 1956 added introduction.

Pauling, Linus, *Vitamin C and the Common Cold*; 1970.

——— *Vitamin C and the Common Cold and Flu*; 1976.

Penrose, Roger, *The Emperor's New Mind*; Oxford University Press; 1989.

——— *Shadows of the Mind*; Oxford University Press; 1994.

Pinker, Steven, *How the Mind Works*; W.W. Norton and Co, 1997.

Pohl, Frederik, *Beyond the Blue Event Horizon*; Del Rey, 1980.

——— *The Coming of the Quantum Cats*; Ballantine, 1986

Robinson, Kim Stanley, *Blue Mars*; Bantam Books, 1996.

——— *Green Mars*; Bantam Books, 1994.

——— *Red Mars*; Bantam Books, 1993.

Rorvik, David M., *In His Image: The Cloning of a Man*; Lippincott, 1978.

Russell, Mary Doria, *The Sparrow*; Ballantine Books, 1996.

Sagan, Carl, *Contact*; Simon and Schuster, 1985.

Schrödinger, Erwin, *What Is Life?*; 1944.

Sheffield, Charles, "A Certain Place in History"; *Galaxy*, 1977.

———— *Aftermath*; Bantam Books, 1998.

———— *Cold as Ice*; Tor Books, 1992.

———— *The Compleat McAndrew*; Baen Books, 2000.

———— "The Hidden Matter of McAndrew"; *Analog*, 1992, "All the Colors of the Vacuum"; *Analog*, 1981. Collected in *The Compleat McAndrew*; Baen Books, 2000.

———— *One Man's Universe*; Tor Books, 1983; published in expanded form as *The Compleat McAndrew*; Baen Books, 2000.

———— *Proteus Unbound*; Del Rey Books, 1989; collected in *Proteus Combined*; Baen Books, 1994.

———— *Sight of Proteus*; Ace Books, 1978; collected in *Proteus Combined*; Baen Books, 1994.

———— *Tomorrow and Tomorrow*; Bantam Books, 1997.

———— *The Web Between the Worlds*; Ace Books, 1979.

Sheffield, Charles and Jerry Pournelle, *Higher Education*; Tor Books, 1996.

Shelley, Mary, *Frankenstein, or the Modern Prometheus*; 1818.

Sterling, Bruce, *Holy Fire*; Bantam Books, 1996.

Sturgeon, Theodore, *More Than Human*; Ballantine Books, 1953.

Swift, Jonathan, *Gulliver's Travels*; 1726.

Tipler, Frank J., "The Omega Point as *Eschaton*: Answers to Pannenberg's Questions for Scientists"; *Zygon*, Volume 24, June, 1989.

———— *The Physics of Immortality*; Doubleday, 1994.

———— "The Twin Paradox for Constant Acceleration Space Craft"; *Journal of the British Interplanetary Society*, Volume 49, 1996.

Varley, John, *The Ophiuchi Hotline*; Dial, 1977

———— *Titan*; Berkley-Putnam, 1979.

———— *Wizard*; Berkley-Putnam, 1980.

Velikovsky, I., *Worlds in Collision*; Doubleday, 1950 (reprinted by Dell, 1965). For a refutation of Velikovsky's work on astronomical and physical grounds, see for example the essay "Venus and Dr. Velikovsky," in *Broca's Brain* by Carl Sagan; Ballantine Books, 1980.

Verne, Jules, *Five Weeks in a Balloon*; 1863.

——— *From the Earth to the Moon*; 1865.

——— *Journey to the Center of the Earth*; 1863.

——— *Twenty Thousand Leagues Beneath the Sea*; 1870.

Watson, James, *The Double Helix*; Atheneum, 1968.

Weinberg, Steven, *Dreams of a Final Theory: The Scientist's Search for the Ultimate Laws of Nature*; Basic Books, 1992.

——— *The First Three Minutes*; Basic Books, 1977.

Wells, H.G., *The Invisible Man: A Grotesque Romance*; 1897.

——— *The Time Machine*; 1895.

——— *The War of the Worlds*; 1898.

White, T.H., *The Once and Future King*; Putnam, 1958.

Index

A

absolute zero 40, 41, 43, 44, 45, 48, 52, 53, 81
Adams, John Couch 173
adiabatic demagnetization 44, 45
Adleman, Leonard 252
AIDS 135, 294
air resistance 194, 269
aliens 7, 8, 9, 10, 11, 103, 115, 120, 182, 237, 328, 329
Allen, Roger 3
Alpha Centauri 9, 66, 182, 207, 208, 209, 230
Alvarez, Luis 303, 304
Alvarez, Walter 303, 304
Amalthea 165, 167, 168, 185
amino acids 10, 127, 137
Ampère, André Marie 46
anaerobic life-forms 167
analytical engine 247
anchored satellite 215
Anderson, Philip 22, 143
antibiotics 1, 147
antimatter rockets 192, 202
antiproton 36, 202, 203, 204
archaic life 140
argon 43, 57, 103
Ariel 172, 173, 175, 187, 188
Arp, Halton 322
Arrhenius, Svante August 316
Artificial Intelligence 261, 312
Artificial Life 16, 247, 264, 266
artificial satellites 2

Asaph Hall 13, 161
Asteroid Belt 162, 163, 230
astrology 328
atomic number 42, 43, 57, 102, 104
atoms 21, 22, 24, 25, 47, 48, 50, 59, 61, 79, 80, 81,
    82, 85, 89, 98, 99, 101, 102, 103, 106, 107, 110, 113, 114,
    117, 118, 119, 198, 238, 239, 250, 263, 281, 282, 309,
    314, 319, 320
automobile 2, 133, 248, 340

B

Babbage, Charles 247, 258
Baen, Jim *v*
Bandwagoner 15, 17
Bard, the 15
Bards 15, 16
basic computational unit 314
beanstalks 192, 215, 217, 219, 220, 228
Bear, Greg 264
Becquerel, Henri 22
Bednorz, Georg 52
bifurcation 273
Big Bang 71, 78, 79, 80, 81, 82, 83, 84, 86, 87, 89, 90, 98,
    322, 323, 324, 325
Big Crunch 85, 89, 90, 94, 95, 96, 99
biological computers 251, 258, 265
black hole(s) 1, 34, 38, 67, 68, 69, 70, 71, 72, 73, 78, 85,
    92, 93, 100, 244, 285
Bohr, Niels 24, 27, 311
boiling under reduced pressure 39, 40
Bose-Einstein statistics 50
bosons 50, 51, 53
bradyons 242, 243
Brin, David 73
Broglie, Louis de 24
brown dwarf 180
buckyballs 118, 119
Burbidge, Geoffrey 322
Bussard Ramjet 192, 205, 209, 210, 211, 212

C

Cairns-Smith, A.G. 139
Callisto 165, 167, 168, 171, 185, 228

caloric value 107
canals of Mars 159, 169, 171
Capek, Karel 260
carbon cycle 62
Casimir Effect 213
CAT scans 1
central dogma of molecular biology 127
Chandrasekhar's limit 66
channeling 328
chaos 38, 267, 268, 277, 281, 283, 285, 347
chaos theory 267, 268, 270, 272, 273, 285, 286
chemical propulsion 192
chemical rockets 191, 192, 196, 197, 198
chemosynthesis 139
chromosomes 125, 140, 141, 251
circadian rhythms 157
clairvoyance 328
Clarke, Arthur 213, 214, 325, 341
Clays 139
Clement, Hal 121, 184
Clementine spacecraft 158
clones 124
closed shells 103
Cocconi, Giuseppe 233, 348
codons 127
cold fusion 318, 321, 322
collapsing the wave function 27
cometary reservoir 177
communication over interstellar distances 232
complementary string 254, 255, 257
computer science 21, 312
continuous creation 322, 325
continuous fusion process 210
Conway, John Horton 265
cosmic background radiation 81, 235
cosmological constant 79, 96, 97
couplets 51
critical density 85, 86, 89
cryovolcanoes 176
curvature of space 33, 338
cyborgs 295

D

dark companion 178, 183

dark matter 85
Darwin, Charles 14, 61, 269
Darwin, Erasmus 14
Davy, Humphrey 39
Dawkins, Richard 128, 129, 143, 305
death rays 290
Deimos 13, 161
deoxyribonucleic acid 123
DEUCE 249
deuterium 109, 200, 201, 319, 324
dinosaurs 138, 241, 302, 303, 304, 305
Dione 169, 186, 187
Dirac, Paul 25
dirty snowball 178
diseases from space 315, 347
distant planets 179
divination 328
DNA 1, 10, 123, 124, 125, 126, 127, 128, 130, 133, 134, 135,
    136, 137, 139, 140, 141, 142, 143, 240, 251, 252,
    253, 254, 255, 256, 257, 258, 265, 266, 293, 347
double helix 124, 125, 126, 135, 140, 254, 310, 350
dowsing 328
Drake Equation 234
Drake, Frank 234, 236, 246
Drexler, Eric 263
dualism 22
dust pools 158
dwarf stars 63, 92, 94, 99, 100
dynamic beanstalks 192, 219, 220, 221
Dyson, Freeman 36, 92, 223, 234

E

Earth-crossing asteroids 162
Eckart, Carl 25
Eddington, Arthur 22
Edgeworth-Kuiper Belt 177
effective jet velocity 195
Einstein, Albert 23, 67, 79, 244, 270
EJV 195, 196, 197, 198, 199, 200, 201, 204, 211
EK Belt 177
electrical conduction 48
electromagnetic force 35, 197, 271
electron 22, 26, 28, 34, 35, 36, 48, 50, 51, 53, 54, 80, 82,
    83, 89, 93, 98, 99, 102, 103, 105, 106, 107, 109, 110,

111, 114, 115, 200, 202, 210, 249, 250, 259, 263, 290, 291, 292, 297
electron pairing 51
electron shells 102, 292
electron spin 26, 28
electron volt 110, 249, 250
Enceladus 169, 186, 187
End of Science 339, 340, 347
energy content 85, 107
ENIAC 248
"entangled" particle pairs 29
enzymatic engine 116, 117
enzymes 116, 135
epidemics 316
ergosphere 69, 70
escape velocity 68, 159, 166, 179, 180, 219
eschaton 90, 349
Europa 165, 167, 171, 185
event horizon 68, 69, 73, 348
Everett, Hugh 27
excited state 290, 291
exoskeletons 10
expansion forever 91
expansion of the universe 1, 32, 81, 85, 86, 96, 342
extinction of the dinosaurs 138

F

faith healing 328
false futures 223
faster-than-light 9, 29, 72, 230, 241
faster-than-light transportation 29
Feigenbaum Number 267, 277, 347
Feinberg, Gerald 242
Fermi Paradox 237
Fermi-Dirac statistics 49
fermions 49, 50
Feynman, Richard 36, 212, 258, 263, 341
fifth force 35
finite state automata 1
Fitzgerald-Lorentz contraction 31
Fleischmann, Martin 319
forbidden transition 290, 291
Forward, Robert 208, 213
Fowler, William 63

Fracastoro, Girolamo 317
fractal dimension 283
fractals 279, 280, 283
Frankenstein 13, 14, 15, 349
Franklin, Ben 46
free energy 96, 299, 325
free radicals 144
FTL 9
fullerenes 117, 118, 119, 120
fundamental equation of rocketry 196
future war 287

G

Gaia theory 157
galactic super-clusters 324
galaxies 34, 72, 75, 76, 77, 78, 79, 81, 85, 86, 87, 91, 96,
    97, 140, 203, 232, 322, 323, 324
Galilean satellites 165, 167, 185
game theory 16
Gamow, George 80
Ganymede 165, 167, 171, 185
Gauss, Carl Friedrich 231
Gell-Mann, Murray 36
general theory of relativity 33, 79, 244
genetic code 124, 139, 141, 251, 294
genetic engineering 1
genome 1, 139, 251, 293, 295
genome mapping 1, 251, 295
geostationary orbit 214, 215, 216, 217, 218, 225
geostationary satellite 214
geosynchronous satellite 214
germ theory of disease 317
Glaser, Peter 224
global altruism 307
Golding, William 306
Gould, Stephen Jay 305
gravitational collapse 65
gravitational contraction 60
gravitational force 35, 67, 87, 179, 183, 220
gravitational radius 68, 69
gravity swingbys 192, 205, 207
Great Dark Spot 174
Great Red Spot 164, 166, 169, 174
Great White Spot 170

ground state 213, 290, 291, 292, 297, 298, 299
Guillaume, Charles 323

H

Haldane, J.B.S. 10, 97
Hawking radiative process 71
heavenly funicular 215
Heisenberg, Werner 24
Heisenberg's uncertainty principle 26
helium II 43, 44
helium-3 200, 201
Helmholtz, Hermann von 60
Herschel, William 75, 171
hexing 328
HIV 135
holography 1
Hoyle, Fred 63, 79, 153, 315, 322
Hubble Space Telescope 170
human thought 261, 313, 314
Hutton, James 60, 306
hydroxyl radical 235, 236
Hyperion 169, 186, 187

I

Iapetus 169, 171, 186, 187
immortality 140, 145, 349
immune system 146, 147, 148
Importers 15, 16
Independence Day 8, 328
Indian rope trick 219, 220
inflationary phase 83
information theory 1
inorganic crystals 139
International Space Station 223, 224, 226
Internet 2, 3
interstellar travel 204, 229
introns 127
inverse perturbations 173
Invisible Man 12, 288, 350
Io 165, 166, 167, 185
ion rockets 191, 198
ionization potential 110, 111
island of stability 105

isomers 113
isotopes 109
iterated functions 273, 277, 284

J

Jet Propulsion Lab 3
Joule, James Prescott 40
Jupiter 6, 154, 162, 163, 164, 165, 166, 167, 168, 169, 172,
    174, 176, 179, 180, 181, 183, 185, 190, 201, 205, 207, 222,
    228, 235

K

K/T extinction 303, 305
Kelvin 40, 45, 50, 71, 174, 176, 235, 323
kernel 69, 70, 71, 73, 244
Kidjel Ratio 329, 330
kinematic relativity 85
Kress, Nancy *v*, 149
krypton 43, 57, 103
Kuiper Belt 177

L

L-5 colonies 222, 223, 224
Lagrange, Joseph Louis 222
Lagrange points 222
laser beam propulsion 192, 205, 208
laser-powered rockets 192, 205, 211
lasers 1, 101, 120, 211, 291, 292
latent heat of liquefaction 39
launch loops 192, 220, 221
levitation 55, 328
librations 157
life elsewhere 138
life in the far future 93
life-forms on Mars 6, 342
light-years 8, 9, 18, 27, 29, 75, 77, 78, 178, 184, 200, 209,
    230, 231, 237, 241
linear accelerators 198
load-bearing cable 217
long-period comets 163, 177
Lord Kelvin 40, 60, 61
Lorentz contraction 31
Lovelock, James 121, 157, 305

Lowell, Percival 159, 169
Lucian of Samosata 12
Lunar Prospector spacecraft 158
Lunik III 158

M

mad cow disease 136
magic numbers 104, 105
magnetic levitation 55
main sequence 62
major histocompatibility complex 294
Mandelbrot Set 279, 281, 282, 283
many-worlds theory 133
Marco Polo 12
Margulis, Lynn 157, 307
Mars atmosphere 160
Mars Sampler spacecraft 140
mass drivers 191, 197, 198
mass-luminosity law 62
matter/antimatter annihilation 203
Mayr, Ernst 22
Meissner Effect 55
Mendeleyev, Dmitri 42
Mercury (planet) 151, 152, 153, 180, 181, 208, 228
mercury (element) 57, 111
Messier Catalog 76
meteorology 21
microtubules 314, 315
microwave region 234, 235, 323
Milky Way 75, 76
Mimas 169, 171, 186, 187
Minsky, Marvin 21, 262, 312, 313, 348
Miranda 173, 187, 188, 348
mirrormatter 202
missing matter 84, 86, 87, 91, 322
mitochondria 125
mixed states 26, 27, 259
Moon 1, 6, 12, 13, 14, 15, 138, 139, 151, 152, 156, 157,
    158, 159, 165, 166, 167, 171, 173, 176, 177, 179, 186,
    187, 189, 194, 218, 222, 224, 229, 230, 305, 349
Moravec, Hans 219, 262
Morrison, Philip 233, 236
Moseley, Henry 42
Müller, Alex 52

muon 36, 49, 107, 202, 204
muonium 107, 122
mutation 126, 129, 130, 131, 132
mystery of sex 128

N

N-body problem 183
nanotechnology 119, 247, 263, 264, 265
National Institutes of Health 3
natural selection 129, 131, 269, 270
Nemesis 178
neon 43, 57, 63, 64, 65, 103
Neptune 168, 173, 174, 175, 176, 177, 189, 200, 201, 208, 228
Nereid 175, 189
neurons 239, 251, 314
neutral hydrogen 110, 210, 235
neutrino 36, 49, 82, 83, 85, 86, 87, 91, 98, 99, 202, 204
neutron 34, 35, 36, 37, 49, 65, 66, 67, 72, 80, 82, 83, 92, 93, 100, 104, 105, 106, 109, 110, 111, 201, 207, 250, 263, 299, 320, 321
neutron excess 105
Neutron star 65, 66, 67, 72, 92, 93, 100, 207, 250
New Age 3, 329
Newton, Isaac 38, 183, 342
Niven, Larry 73, 121, 348
noble gases 43, 103
non-zero cosmological constant 97
nonlocality of the universe 28
nonprescription drugs 144
nonsynchronous skyhook 219
nuclear bombs 1
nuclear energy 1
nuclear laser 297, 299
nuclear reactor rockets 191, 198, 199
nuclear Test Ban Treaty 200
nuclear war 287, 296, 304, 308
nuclear winter 304
nucleic acids 123
nucleus 23, 34, 35, 36, 42, 62, 83, 102, 103, 104, 105, 106, 107, 109, 111, 114, 124, 125, 201, 210, 290, 293, 297, 298, 299

O

Oberon 172, 173, 187, 188
Oersted, Hans Christian 46
Ohm, Georg Simon 46
one-way membrane 68
O'Neill, Gerard 222
Onnes, Kamerlingh 43, 44, 45, 46, 54
Oort Cloud 177, 178, 316
Oort, Jan 177
open universe 91, 92, 93, 95, 100, 346
orbital tower 215
organ transplants 146
organic chemistry 25, 114
origin of life 133, 137, 138, 346

P

palmistry 328
panspermia 10, 316
parallel operation 250, 259
parsecs 18
Pathfinder lander 160
pathological curves 283
Pauli Exclusion Principle 49, 50
Pauli, Wolfgang 36
Pauling, Linus 25, 309, 310, 311
pendulum 273, 280, 281, 284, 285
Penrose, Roger 21, 69, 266, 312, 348
periodic 42, 104, 178, 181, 273, 284, 285
personal computers 1, 289
personal weapon 292
Phobos 13, 161
Phoebe 169, 170, 186, 187
photoelectric effect 23, 343
photon rockets 192, 204
photons 23, 24, 36, 50, 288
physics 1, 11, 21, 23, 37, 39, 45, 51, 59, 60, 72, 75, 79,
    87, 91, 101, 103, 111, 123, 184, 198, 203, 204, 205, 209, 212,
    230, 242, 244, 247, 249, 263, 278, 281, 295, 303, 313,
    315, 320, 322, 323, 331, 339, 340, 341, 346, 347, 349
Pilkington, Ace FM-v
pions 36, 37, 202, 203, 204
Planck length 38, 244
Planck, Max 23, 340

Planck time 83, 84
Planck's constant 23, 24, 249
Planet X 177
plasma cosmology 323, 324, 325
plate tectonics 156, 157
Pluto 151, 164, 176, 177, 228, 229
Pohl, Frederik 27, 73, 348
Pons, Stanley 320
positron 34, 35, 82, 89, 93, 98, 99, 107, 202
prescience 328
presynaptic vesicular grid 314
Prigogine, Ilya 22
primordial methane 157
prion protein 136
prions 16, 133, 137
problem of decoherence 260
Project Cyclops 236
Project Daedalus 200, 201
Project Orion 199, 200
properties of materials 21
proton decay 92, 93
protons 24, 34, 35, 37, 42, 49, 65, 82, 92, 104, 105, 106,
     109, 111, 202, 299
Proxima Centauri 182
psychics 328
psychokinesis 328
pulsar 65, 236
pulsed fission rockets 191, 199

Q

quanta 23, 24
quantum chromodynamics 36, 37, 38
quantum computers 258, 259, 262
quantum computing 251
quantum electrodynamics 37, 38
quantum indeterminacy 25, 26, 27
quantum paradoxes 25
quantum teleportation 27, 31, 33, 241, 244, 346
quantum theory 1, 5, 21, 23, 25, 26, 27, 28, 29, 34, 36, 38,
     40, 48, 49, 70, 103, 212, 258, 259, 268, 309, 313, 342
quark(s) 16, 19, 36, 37, 72, 83, 106
quasars 16, 72, 78, 322
quasi-stellar radio source 78

R

R.U.R. 260
radio noise 165
radioactivity 18, 22, 23, 25, 129, 297, 299, 304
radon 42, 43, 57, 103
Ram Augmented Interstellar Rocket (RAIR) 192, 205, 212
reaction mass 191, 192, 193, 195, 196, 197, 198, 199, 203, 205, 210, 211, 213
recession of the galaxies 77, 322, 323
rectenna 224
red giant 66, 88, 99, 100
redshift-distance correlation 322
refractive index 12, 288
regolith 161
reincarnation 328
relativity 1, 23, 25, 29, 30, 31, 32, 33, 34, 67, 342, 346
resistance 18, 30, 44, 45, 47, 48, 50, 194, 269, 294
resonance lock 152
resonances 299
retroviruses 1, 16, 135
Rhea 169, 186, 187
ribonucleic acid 126
ribosomes 127, 135
ribozymes 135
Rigel 7, 8, 9, 11, 182, 183
ring singularity 70, 244
ring system 168
rings of Saturn 168, 170, 172
RNA 10, 126, 127, 133, 135, 136, 137, 139, 140
Robinson, Kim Stanley 160, 348
robots 247, 260, 262, 296, 346
rocket spaceships 191, 192, 204
rocketless spaceships 191
Röntgen, Wilhelm 22
room-temperature superconductors 57
rotating beanstalk 219
Russell, Bertrand 30, 315
Rutherford, Ernest 23

S

Saturn 163, 164, 168, 169, 170, 171, 172, 174, 176, 178, 179, 180, 186, 201, 208, 228
scaling 281, 282, 283

Schrödinger, Erwin 24, 124, 348
Schrödinger's Cat 25, 27, 347
Schwarzschild black hole 67, 69, 70, 73
Schwinger, Julian 36
science fiction 2, 3, 5, 6, 7, 8, 9, 11, 12, 13, 16, 17, 25,
    27, 35, 38, 69, 72, 84, 86, 88, 101, 102, 105, 107, 110,
    115, 116, 121, 123, 133, 143, 145, 151, 153, 159, 162,
    173, 176, 179, 192, 195, 197, 209, 213, 216, 229, 230,
    238, 239, 248, 260, 262, 265, 286, 288, 290, 295, 328,
    329, 331, 340, 343, 344
scientific heresies 301
scrapie 136
Search for Extraterrestrial Intelligence (see: SETI) 230, 237,
    238, 347
seer 15, 17
self-regulating organism 305, 307
self-replicating entities 139
sensitive 15, 17
serial operation 250
SETI (see: Search for Extraterrestrial Life) 230, 234, 236, 237,
    238, 240, 348
sexual reproduction 128, 130, 132, 133, 140
shadow matter 87
Shelley, Mary 13, 14, 15
SI 195, 196, 197
sick curves 280
sidereal period 154
silico-organic 115
Sirius 182, 183
sister planet 153, 156, 159
skyhook 215, 219
Sojourner 160
solar power satellites 223
solar radiation pressure 208
solar sails 192, 205, 207
solar system 78, 83, 88, 98, 99, 138, 139, 151, 154, 155,
    157, 161, 162, 163, 164, 167, 168, 169, 171, 173, 176,
    177, 178, 179, 181, 182, 183, 193, 198, 199, 201, 205,
    206, 217, 218, 228, 229, 230, 231, 233, 235, 292,
    316, 342
solid metallic hydrogen 217
solid state physics 1
space colonies 178, 221, 223
space fountain 192, 219
spaceflight 5, 191, 194, 196

spacetime 31, 34, 95, 212, 244
special relativity 29, 30, 31, 241, 242, 342
specific impulse 195
speed of lightness 333, 334, 338
spin state 26, 28, 259
spinning black hole 69, 70, 71
spontaneous emission 235, 290, 291, 292
SPS 224, 225, 226
stars 13, 22, 24, 59, 61, 62, 63, 64, 65, 66, 75, 76, 77, 78,
    79, 80, 81, 82, 85, 92, 93, 94, 99, 100, 101, 118, 172,
    178, 179, 180, 181, 182, 183, 184, 190, 203, 207, 209,
    229, 230, 231, 232, 233, 234, 235, 238, 240, 243, 246,
    295, 319, 323, 328, 331, 345
static universe 79
statistics 49, 50
steady state theory 322
stellar endings 64
stellar nucleosynthesis 63, 64, 81
stellar occultation 175
Stickney 161
stimulated emission 291, 292
stimulated nuclear emission 299
strange attractors 284, 285
strong force 35
subatomic structure 1
super-soldier 295
superconductivity 39, 45, 46, 48, 49, 50, 51, 52, 53, 54, 321
supernova 65, 66, 71, 81, 182, 298, 304, 323
supersonic aircraft 1
superstring 16, 37, 38, 87
supersymmetry 83, 86, 87
surface gravity 180, 195
surface of infinite red shift 68, 69

T

tachyons 242, 243, 244
taper factor 215, 216, 217, 219, 228
telepathy 328
telephones 1
teleportation 27, 31, 33, 241, 244, 328, 346
television 1, 101, 235
telomerase 141, 142
telomere 140, 141, 143
temperature 39, 40, 41, 42, 43, 44, 45, 46, 47, 48, 49, 50,

51, 52, 53, 54, 55, 56, 57, 59, 63, 64, 65, 66, 71, 80,
81, 82, 83, 84, 92, 94, 106, 108, 115, 116, 117, 139,
155, 156, 160, 170, 174, 176, 180, 198, 199, 200, 201,
210, 217, 235, 306, 319, 320, 321, 323, 326
terrestrial microwave window 235
Tethys 169, 186, 187
Theory of Everything 35, 38, 339, 340
Theory of Relativity 23, 29, 30, 31, 33, 34, 79, 240, 244,
269, 343, 345, 348
thermalization 323, 324
Thomson, J.J. 22, 34, 47, 342
Thomson, William 40
thought experiment 28
time dilation 30, 33, 202, 240
tissue engineering 145, 147, 148, 149
Titan 6, 167, 169, 170, 171, 186, 187, 228, 349
Titania 172, 173, 187, 188
Tombaugh, Clyde 176
Tomonaga, Sinitiro 36
transistors 1, 260
transmission of electrical power 56
transuranic elements 104, 106, 107
trapping surface 68
Traveling Salesman Problem 252, 258
Triton 175, 176, 189
tubulin dimers 314
Turbulence 278, 279
two-body problem 183

U

ultraviolet light 8, 11
Umbriel 172, 173, 175, 187, 188
uniformitarianism 60
universality 277, 279, 283
unstable solution 276
Uranus 168, 171, 172, 173, 174, 175, 187, 201, 208, 228,
342
Ussher, Archbishop of Armagh 59

V

vacuum energy 38, 192, 205, 212, 213
vacuum energy drive 192, 205, 212, 213
Vallis Marineris 160

Velikovsky, Immanuel 153
Venus 151, 153, 154, 155, 156, 159, 171, 228, 231, 349
Verne, Jules 14, 15
Verrier, Urbain Le 173
Viking Lander 160
viruses 123, 133, 134, 135, 293, 294, 315, 316
viscosity 44
vital force 113
vitalism 113, 115
Vitamin C 144, 308, 309, 310, 311
"Voyage to Laputa" 13
Voyager 2 169, 171, 172, 173, 174, 175

W

weak force 35
web sites 3
Wheeler, John 27, 68, 70
white dwarf 66, 88, 183
Wickramasinghe, Chandra 315, 347
Wigner, Eugene 26
world-builders 15, 17, 154
wormholes 243, 244

X

X-rays 22, 300
xenon 43, 57, 103

Z

Zweig, George 36